Organic Chemistry:
A Guided Inquiry

Suzanne M. Ruder, PhD
Virginia Commonwealth University

POGIL Project
Director: Richard Moog
Associate Director: Marcy Dubroff
Publication Liaison: Sarah Rathmell

ISBN-13: 978-1-119-23460-9
ISBN: 1-119-23460-3

SKY10029107_081321

To the Student

Process Oriented Guided Inquiry Learning (POGIL) is a method of instruction where each student takes an active role in the classroom. The activities contained in this collection are specially designed guided inquiry activities intended for the student to complete during class while working with a small group of peers. Each activity introduces essential organic chemistry content in a model that contains examples, experimental data, reactions, or other important information. Each model is followed by a series of questions designed to lead the student through the thought processes that will result in the development of critical organic chemistry concepts. At the end of each activity are additional questions, which will generally be completed outside of class time and are more similar to questions that might appear on tests. Before each class, students should ensure that they are familiar with the prior knowledge that is listed at the beginning of every activity.

The POGIL process of learning helps the student understand the content as well as develop critical process skills (such as critical thinking, problem solving and teamwork) that are important in the workplace. Using this method, students will learn to apply scientific concepts, analyze and evaluate scientific information, and communicate these ideas to others. The role of the instructor (and any teaching assistants) is to guide students through this process, facilitate discussion, and provide a context and conceptual framework for the course material. The student will learn to ask questions, read and evaluate information, assess whether or not they understand new ideas, and collaborate with others to develop new knowledge.

For many students, this may be the first time experiencing this type of learning environment. While it may at first seem confusing, ultimately students become familiar with the process of exploring information, discovering new concepts, and solving problems in the context of a variety of perspectives and approaches through peer collaboration. It is important to note that work done in the classroom must be followed by active work outside the classroom. This includes reading a textbook to provide further detail and context, reviewing the answers to these activities and confirming an understanding of the topics by application to homework problems. The concepts and ideas that are initially developed using the activities in class will only become well-developed knowledge through repeated experience and exposure to the content materials

To the Instructor

The POGIL activities in this collection follow the constructivist learning cycle with an exploration, concept invention and application cycle within each model, where possible. Additionally, the materials include cues for team collaboration and self-assessment. In the exploration phase of each activity, students explore a model, which may be an example, figure, experimental data, or series of reactions. Directed questions encourage students to explore the model, especially the fundamental relationships or concepts embedded in it. Concept development is brought about by convergent questions, which guide students to make connections and form hypotheses, ultimately leading them toward the understanding of the desired concept. Term introduction typically occurs at this phase, to help students identify or label the newly developed concept. Students are often asked to clearly articulate and explain the concept. In addition to developing communication skills, clear expression of the concept is essential for a true understanding. In the application phase of the activity, students are asked to apply the concept to new situations. These exercises help to build students' confidence and strengthen their understanding. Application questions that include more advanced problems are found at the end of each activity (additional questions). These problems require higher level thinking skills such as synthesis or analysis, and may involve a multi-step approach integrating several concepts or skills.

The activities in this collection were designed to be used in the classroom and have been implemented in settings with class periods ranging from 50 to 75 minutes, and class sizes ranging from 20-250 students. Students should work through the activities in groups of 3-4 with the instructor and any teaching assistants actively facilitating student engagement and learning during class. The activities listed in the Table of Contents are in an order consistent with textbooks that follow a functional group approach. Each activity lists the assumed prerequisite knowledge to allow the most flexibility on the part of the instructor for sequencing the use of the activities. The materials are not meant to replace a textbook, rather they are designed to enhance the learning experience in the lecture component of the course. Due to the guided inquiry design of the activities, it is recommended that students complete the activity first and subsequently read the corresponding sections of the textbook. Instructors should also assign problems from the text or online homework system as part of homework assignments.

Each instructor will naturally implement the activities in ways that are unique to that individual and institution. The author and other faculty that classroom tested these activities found implementation that included mini-lectures and whole-class discussions, in addition to the small group work, to be very effective. Students may initially move slowly in this learning process, but eventually they become more adept as activity usage builds and they gain insights into learning strategies and confidence in themselves and teammates. For this reason, it is important to use activities on a regular basis (daily, weekly, or biweekly) so that students become accustomed to this learning environment and their role in it. The facilitation guide for instructors contains tips on facilitation of the activities, and questions that are useful to have students report out, or for whole class discussion. Some implementation strategies include assigning initial directed questions in the activity as pre-class work to be discussed at the beginning of class, using the application questions as homework, providing mini-lectures to introduce background material or to address common sticking points, using whole-class discussions to compare student responses and elaborate on ideas, and providing a summary at the end of the class to highlight the learning objectives addressed. It is also recommend that instructors attend a POGIL workshop to gain a deeper understanding of the philosophy and facilitation of activities.

It is important that students get some feedback on content mastery gained through use of the activities– either through classroom summaries, graded questions, or facilitator review of student work as it unfolds in the classroom. Instructors do not typically grade the activities, but may focus on particular application problems or

a small subset of key questions, particularly those that are summative in nature. These questions can be graded individually, as homework, or for the group, either as a clicker question, electronic document or a paper copy. It is also important to explicitly address process skill mastery – which entails some form of evaluating student team efforts, communication skills, information processing, self-assessment (metacognition), and critical thinking as well as the more traditional problem solving strategies. Both process skill and content learning objectives are listed at the start of each activity.

The facilitation guide for these activities includes the following information: (1) Type of Activity; (2) Prior Knowledge; (3) Learning Objectives (Content and Process); (4) Facilitation Notes; and (5) Additional Concepts. Student activities include the prior knowledge and learning objectives in order to provide them with specific information about what they should already know and what they will learn after completing the activity. The learning objectives have details for both content and process skill goals that help explain the goal within the context of the specific activity. The facilitation notes include suggestions for reporting out for each model and some guidance about which concepts that students typically get stuck on. The section on additional concepts is where concepts that are related but are not included in the activity are mentioned. Generally, these concepts can be discussed in a mini-lecture or assigned as additional homework. Concepts not included in an activity are usually ones that are not critical to the understanding of a key foundational concept, and can thus be easily added in a short mini-lecture. Alternatively, a mini-activity could be used to work through some of these additional concepts. In a mini-activity, a model with some questions is presented in class on a slide. Usually miniactivities are 1-2 slides, and cover a single topic. The questions can be discussed as a whole class discussion or some can be presented as clicker questions.

About the Author

Dr. Suzanne Ruder is an Associate Professor of Chemistry at Virginia Commonwealth University. Dr. Ruder earned a B.A. in chemistry from the College of St. Benedict, a Ph.D. degree in organic chemistry from Washington State University, and completed a post-doctoral position at Brown University. Her research focuses on training teaching assistants, designing instructional materials and developing methods to assess process skills in the active learning classroom. Dr. Ruder teaches organic chemistry at VCU using active learning methods in large classes (up to 250 students). In addition to authoring this set of POGIL activities for organic chemistry, she has led seminars and workshops about active learning throughout the United States and Australia. She has been involved with the POGIL project since 2003 in a variety of ways including facilitating workshops, training workshop facilitators and serving on the Steering Committee.

Acknowledgements

Special thanks go to the POGIL project, specifically Rick Moog, Marcy Dubroff and all the POGIL staff through the years, for their help and guidance in making this whole endeavor possible. Their guidance and leadership of the POGIL Project has improved access to evidence-based teaching for students and instructors throughout the world. Thanks also to the faculty who have used these activities in their own classrooms and provided feedback to continue to improve the activities. I also wish to thank my many students and teaching assistants who provided their insights into the structure of the activities and helped improve the materials. Finally, I would like to thank my husband Alan and daughters Nicole, Kathryn and Rebekah, for their endless encouragement and support.

Table of Contents

These activities were written to cover most of the important concepts for a two semester organic chemistry sequence. The activities are grouped into organic 1 and organic 2, although that might vary from class to class depending on the textbook used. The order of activities most closely follows the textbook by Wade. Books by Smith and Carey/Gulianno are structured in a similar order. Some concepts do not have an activity, particularly if the concept is of narrow focus. The following are some ideas for introducing additional concepts that do not have an activity.

- Assign the topic as homework/reading outside of class.
- Mini-lecture on the concept.
- Prepare a "mini-activity" on the concept to be done in groups during class. Usually a mini-activity consists of one model and questions on a single slide.

Class Activity 1

Drawing Organic Structures

Prior Knowledge:
Before beginning this activity, students should be familiar with the following concepts:
- Molecular formula
- Lewis structures
- Bonding

Learning Objectives
Content Learning Objectives:
After completing this activity, students should be able to:
- Articulate the difference between a Lewis structure, condensed structure and line-angle structure.
- Convert organic structures between the Lewis, condensed and line-angle forms.

Process Objectives:
- Teamwork. Students interact with others in the group to draw organic structures in different ways and discuss group reflections to submit at the end of the class.
- Information Processing. Student interpret information about condensed, line-angle and Lewis structures, and transform one form into another.

Class Activity 1

Drawing Organic Structures

Model 1: Drawing Organic Structures
Organic structures can be drawn in different ways for the same compound. The three main ways of representing organic structures are shown below.

	Molecular Formula	Lewis	Condensed	Line-angle (skeletal)
1	C_3H_7Br		$CH_3CHBrCH_3$ or CH_3CHCH_3 with Br	
2	$C_4H_{10}O$		$CH_3(CH_2)_3OH$	
3	C_3H_6O		CH_3COCH_3 or CH_3CCH_3 with O	
4	C_8H_{14}		$C_5H_9CH_2CHCH_2$ or $C_5H_9CH_2CH=CH_2$	

Questions:

1. Consider the <u>Lewis</u> structures shown in Model 1:
   (a) What is the molecular formula for the compound shown on line 1?
   (b) In the Lewis structure on line 1, is every atom in the formula represented (i.e. written out)?

(c)   In the Lewis structure, how are bonding electrons represented? How are non-bonding electrons represented?

(d)   In the Lewis structure, are all bonds to every atom represented?

(e)   For the remaining three lines in Model 1, compare the formulas to the Lewis structures. Discuss in your group whether your answers to #1b, #1c, and #1d are consistent with these remaining structures.

2.   Consider the <u>condensed</u> structures shown in Model 1:
   (a)   In the condensed structure, is every atom from the formula represented?

   (b)   In the condensed structure, how are bonding electrons represented?

How are non-bonding electrons represented?

   (c)   In the condensed structure, are all bonds to every atom shown?

   (d)   In the condensed structures reading left to right, which atom always appears on the left?

Which atom always follows the atom on the left?

What atoms can be drawn above or below the line of atoms in the condensed structure?

   (e)   In entry 4 in the table in Model 1, how is the five-membered ring represented for the condensed structure?

   (f)   Based on the above answers, summarize how condensed structures differ from Lewis structures.

   (g)   Once your group has agreed on the above questions, formulate a plan for converting Lewis structures to condensed structures.

   (h)   Convert the following Lewis structure to a condensed structure.

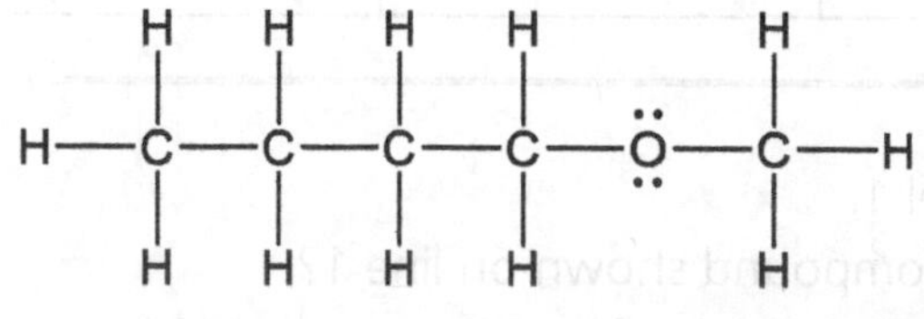

3. Consider the <u>line-angle</u> (skeletal) structures shown in Model 1:
    (a)   In the line-angle structure, is every atom from the formula represented?

    (b)   In the line-angle structure, how are bonding electrons represented?

          How are non-bonding electrons represented?

    (c)   In the line-angle structure, are all bonds to every atom shown?

    (d)   Which atoms are not shown in line-angle structures that are present in the Lewis
          structures?

    (e)   How are carbon atoms represented in line-angle structures?

    (f)   How are heteroatoms (atoms other than C and H) depicted in line-angle structures?

    (g)   For each entry in the table, count the number of carbons in the Lewis structure and
          compare to the condensed and line-angle structures. Check that the number of carbons
          in each structure is the same as the molecular formula provided for each compound.

    (h)   Once your group has agreed on the above questions, formulate a plan for converting
          Lewis structures to line-angle structures.

    (i)   Convert the following Lewis structure to a line-angle structure.

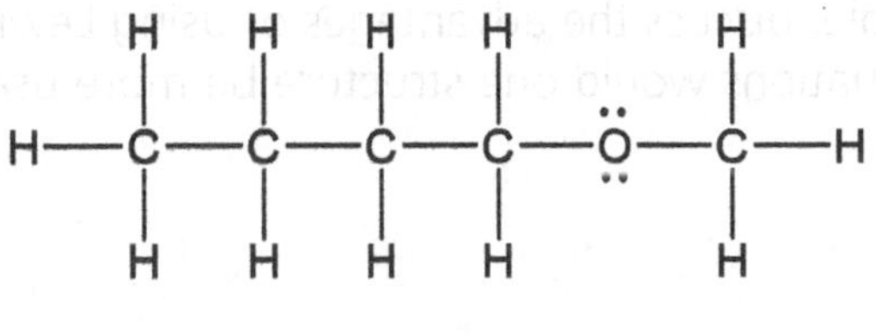

$$
\begin{array}{ccccccccccc}
& & H & & H & & H & & H & & & & H \\
& & | & & | & & | & & | & & & & | \\
H & - & C & - & C & - & C & - & C & - & \ddot{\underset{\cdot\cdot}{O}} & - & C & - & H \\
& & | & & | & & | & & | & & & & | \\
& & H & & H & & H & & H & & & & H
\end{array}
$$

4. Draw the missing structures in the following table.

Lewis	Condensed	Line-angle
(Lewis structure of 4-carbon chain with OH on third carbon)		
		(line-angle structure with COOH and Br)
	CH₃CH₂CH(OH)CH(CH₃)₂	
(Lewis structure of benzene ring with O linked to vinyl group)		

5. Once your group has completed the above table, discuss the advantages of using Lewis, condensed and line-angle structures. What situations would one structure be more useful than another?

Reflection: on a separate sheet of paper.

As a group, describe three concepts your group has learned from this activity and the one most important unanswered question about this activity that remains with your group. Turn this in before leaving class.

Additional Questions:

6. Convert each of the following Lewis structures to condensed structures and line-angle structures.

   (a)

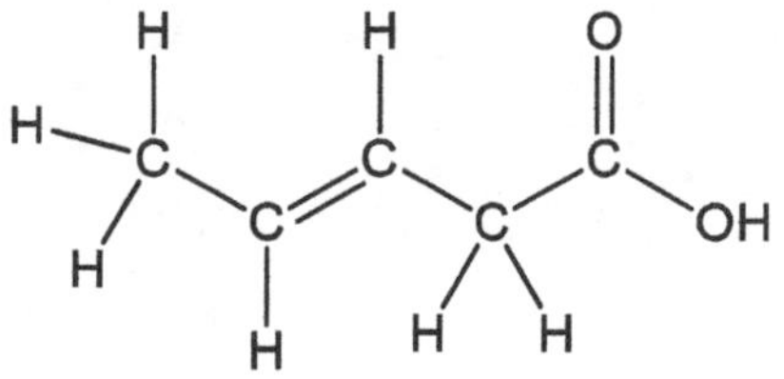

   (b)

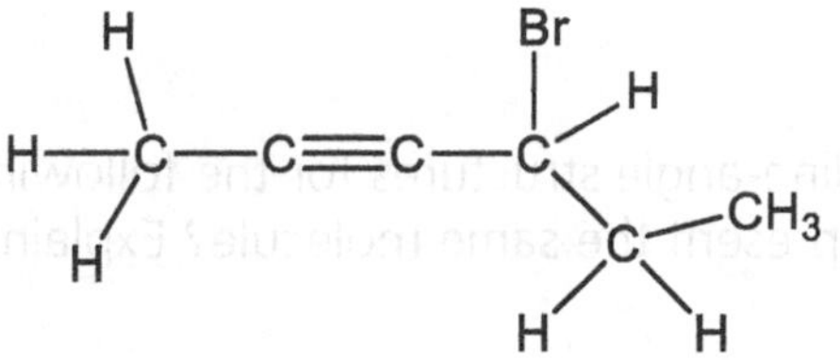

   (c)

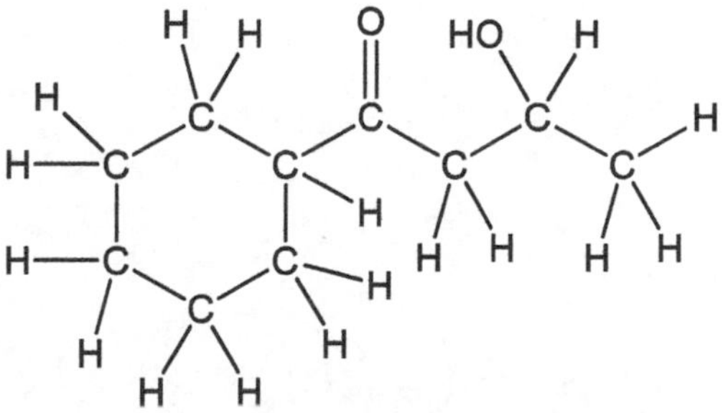

7. Determine the molecular formulas and then write condensed structures for the following line-angle formulas. Do these structures represent the same molecule? Explain.

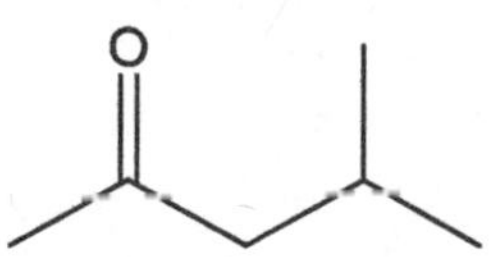

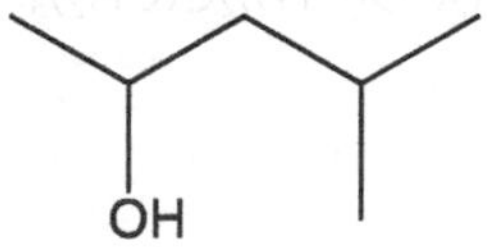

8. Determine the molecular formulas and then write line-angle (skeletal) structures for the following condensed structures. Do these structures represent the same molecule? Explain.

$$CH_3CH_2COCH_2C(CH_3)_3 \quad CH_3COCH_2CH_2C(CH_3)_3 \quad CH_3CH(OH)CH_2CH_2C(CH_3)_3$$

9. Determine the molecular formulas and then draw line-angle structures for the following condensed structures. Do these three structures represent the same molecule? Explain.

(a) $(CH_3)_3CCH_2CH_2COCH_2CHBrCH(CH_3)_2$

(b) $(CH_3)_3CCHCHCHOHCH_2CHBrCH(CH_3)_2$

(c) $(CH_3)_2CHCHBrCH_2CO(CH_2)_2C(CH_3)_3$

10. Determine the molecular formulas and then draw Lewis structures for the following line-angle structures. Do these two structures represent the same molecule? Explain.

11. For each of the following "real-world" compounds, convert each Lewis structure into condensed and line-angle forms.
    (a)   Oxalic acid is a household product used to clean and polish wood and metal items.

    (b)   Glycerol is used in lotions and as a sweetener in foods.

Class Activity 2

Resonance Structures

Prior Knowledge:

Before beginning this activity, students should be familiar with the following concepts:

- Formal Charge
- Lewis structures
- Bonding
- Isomers

Learning Objectives

Content Learning Objectives:

After completing this activity, students should be able to:
- Describe the relationship between resonance structures (that they differ by movement of double bond and lone pair electrons).
- Draw resonance structures of a given compound using curved arrows to show electron movement.
- Determine the major resonance contributor among a series of resonance structures.

Process Objectives:

- Critical Thinking. Students analyze the model of resonance structures to determine how resonance structures differ. They synthesize information from the examples to make conclusions about the purpose of curved arrows and to determine the major resonance contributors in a group.

Class Activity 2

Resonance Structures

Model 1: Resonance Structures for a Carbonyl Group

Questions:

1. (a) Count the electrons in each bond and each lone pair to determine how many total electrons are in structure **1** (_____) and structure **2** (_____).

   Is the total number of electrons in each structure the same? (*Circle one*) yes / no.

   (b) How many single (sigma) bond electrons are in structure **1** (_____) and structure **2** (_____)?

   Is the number of single (sigma) bond electrons the same in each structure? (*Circle one*) yes / no.

   (c) How many lone pair electrons are found in structure **1** (_____) and structure **2** (_____)?

   Is the number of lone pair electrons the same in each structure? (*Circle one*) yes / no.

   (d) How many double (pi) bond electrons are found in structure **1** (_____) and structure **2** (_____)?

   Is the number of double (pi) bond electrons the same in each structure?
   (*Circle one*) yes / no.

   (e) Based on the answers above, describe what is different between structures **1** and **2** (in terms of electrons)?

2. Is the total net charge of structure 1 the same as the total net charge of structure **2** ?
   (*Circle one*) yes / no.

3. As a group discuss what the curved arrow drawn on structure **1** represents.

Model 2: Resonance Structures of Acrylic Acid

Questions:

4. Consider the resonance structures of acrylic acid shown in Model 2:

    (a)   Is the total charge for each resonance form the same? (*Circle one*) yes / no.

    (b)   Is the total number of electrons for each resonance form the same?
          (*Circle one*) yes / no.

    (c)   Have any single bonds been broken to form another resonance form?
          (*Circle one*) yes / no.

    (d)   Are any atoms bonded to a different atom in other resonance forms?
          (*Circle one*) yes / no.

    (e)   Do all atoms have eight electrons or less? (*Circle one*) yes / no.

5. Curved arrows are used to show movement of electron pairs. Consider the curved arrows
   depicted on the structures in Model 2:

    (a)   What does the curved arrow on structure **3** show (be specific)?

    (b)   What does the curved arrow on structure **4** show (be specific)?

    (c)   What do the curved arrows on structure **5** show (be specific)?

    (d)   In converting structure **5** to **6**, it is necessary to move two pairs of electrons. Draw the
          resonance structure that would form if only the <u>one</u> lone pair of electrons on oxygen
          were moved (indicated by the arrow below-remember to only move the pair of electrons
          shown by the arrow). Describe why this is <u>not</u> a valid resonance form.

6. (a)   In structure **3** of Model 2, the double bond electrons move to form a lone pair on <u>oxygen</u>. The arrow below shows the double bond electrons moving to form a lone pair on <u>carbon</u>. Draw the resulting resonance form (make sure to put in charges).

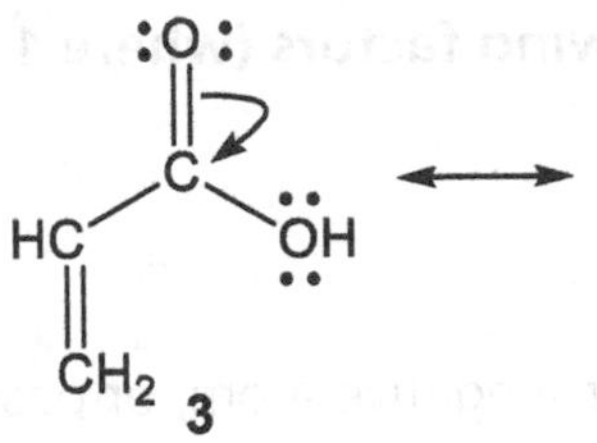

(b)   Although this form is possible, it is <u>not likely</u>. As a group, discuss why the lone pair is more likely to reside on oxygen than on carbon.

7. Follow the resonance structure guidelines developed in question #4 above. For each pair below, determine whether the resonance structure on the right is an acceptable resonance structure of the form on the left. Place an X through any incorrect structures on the right of each pair, and indicate which of the items listed in question 4 were not followed. Once everyone in your group agrees, draw a correct resonance form.

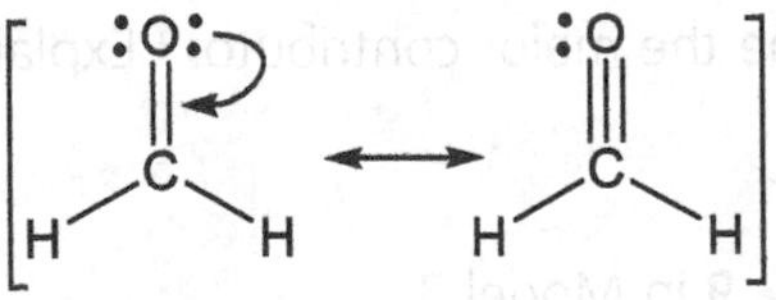

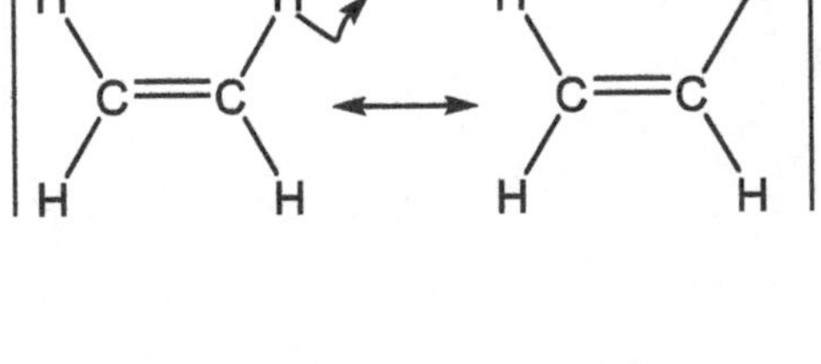

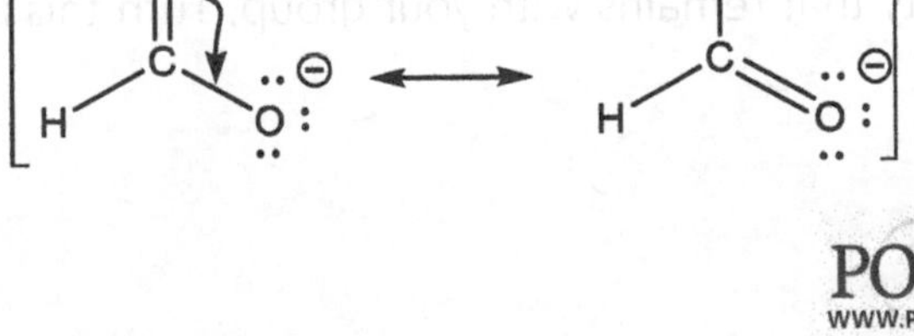

Model 3: Stability of Resonance Forms

Not all resonance forms have the same energy; some forms may be more stable than others.

The major resonance contributor is determined by the following factors (where 1 is more important and 5 is less important):

1. complete octet
2. as many bonds as possible
3. no charge on individual atoms
4. if charges exist, then negative charge resides on more electronegative atom, or positive charge resides on less electronegative atom
5. if charges exist, then small charge separation between charged atoms

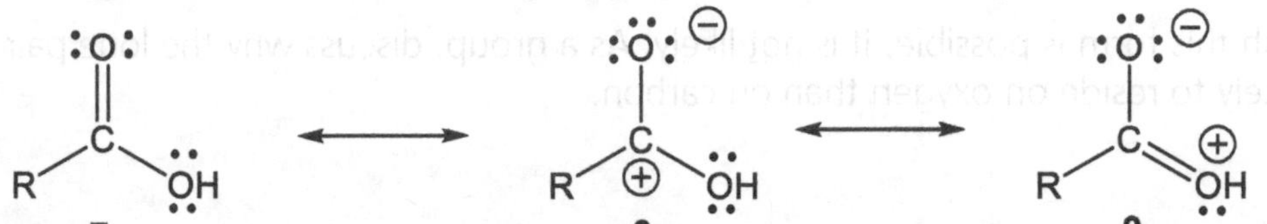

Questions:

8. For the resonance forms shown in Model 3:

    (a) Circle any atoms in Model 3 that do not have complete octets.

    (b) Indicate the total number of bonds for each structure shown in Model 3.

    (c) Which structure has no charges on any individual atoms?

    (d) If charges exist, which atom is most likely to hold the negative charge? (C, H, or O)

    (e) Based on the rules in Model 3, which structure would be the major contributor? Explain.

9. Add curved arrows to show the conversion of **7 → 8** and **8 → 9** in Model 3.

10. For the following structure, use curved arrows to draw **three** additional resonance structures and predict which form, if any, would be the major contributor. (HINT: move double bond or lone pair electrons only and try moving one pair at a time).

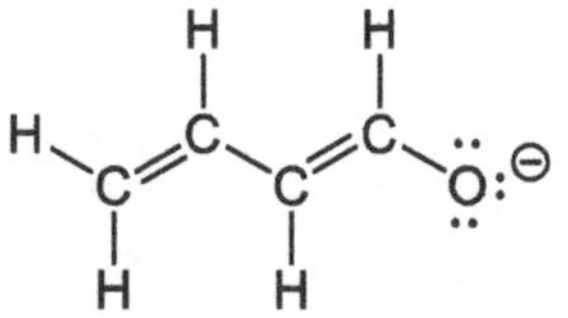

Reflection: on a separate sheet of paper.

As a group, describe three concepts your group has learned from this activity and the one most important unanswered question about this activity that remains with your group. Turn this in before leaving class.

Additional Questions:

11. Use curved arrows to draw any additional resonance structures for the following compounds. Predict which form, if any, would be the major contributor.

(a)

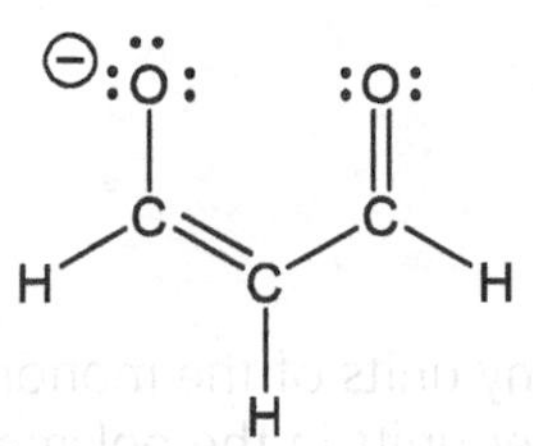

(b)

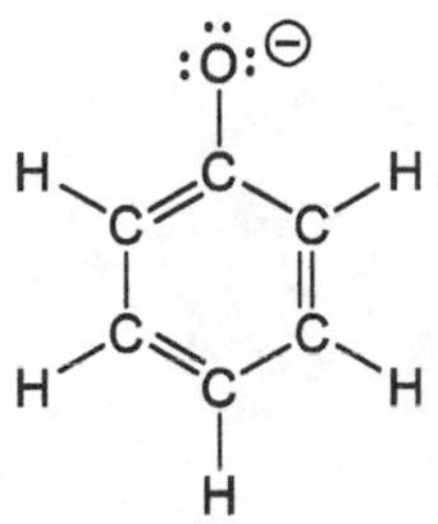

(c)

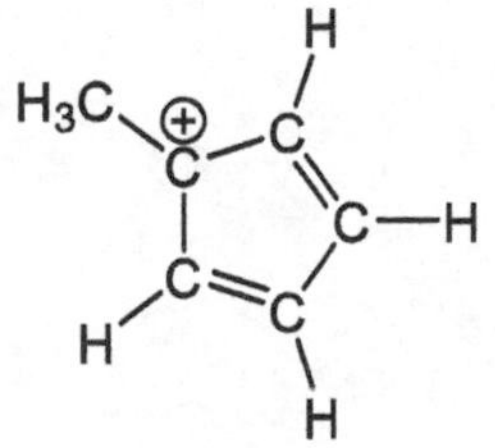

(d)

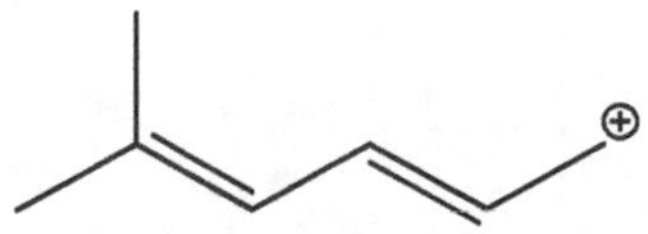

(e)

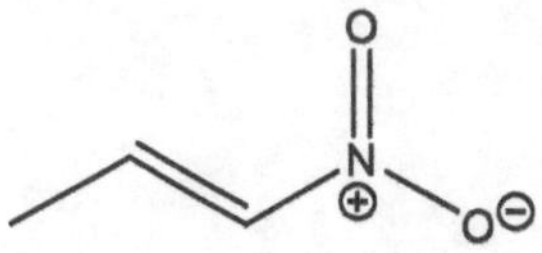

12. Polyacrylic acid shown below is a polymer (many parts) made up of many units of the monomer acrylic acid (from Model 2). Circle the three carbon acrylic acid monomer units in the polymer below. Using resonance structures, explain why carbon 2 of one acrylic acid molecule might be attracted to carbon 3 of another acrylic acid molecule.

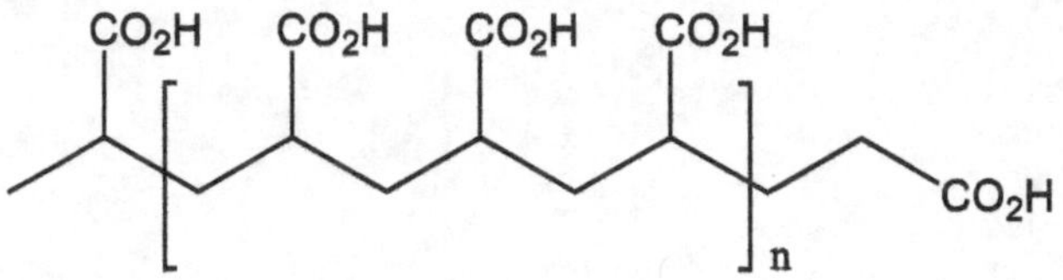

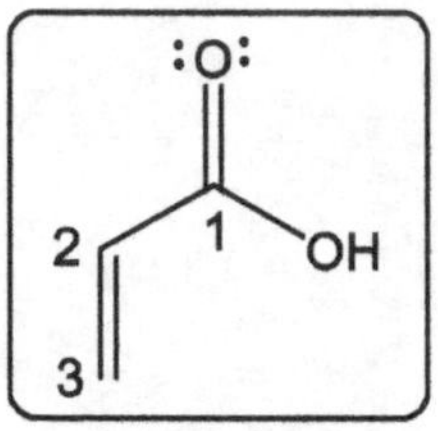

Acrylic Acid

Class Activity 3A

Acids and Bases
Part A: Acids/Bases and pKa Values

Prior Knowledge:
Before beginning this activity, students should be familiar with the following concepts:

- Acid/base definitions
- pKa, Ka, Keq
- Formal charge

Learning Objectives
Content Learning Objectives:
After completing this activity, students should be able to:

- Articulate the differences between the Brönsted-Lowry and the Lewis acid/base definitions.
- Predict acid and base strength based on pKa values and predict the direction of an acid/base reaction.
- Draw curved arrows to depict the mechanism of an acid/base reaction.

Process Objectives:

- Information Processing. Students interpret acid base definitions and pKa values to determine the acid/base roles of each reagent and predict acid/base strength.
- Critical Thinking. Students analyze acid/base reactions to determine a mechanism and to predict the direction of the reaction based on pKa values.

Class Activity 3A

Acids and Bases
Part A: Acids/Bases and pKa Values

Model 1: Acid/Base Definitions

	Brönsted-Lowry	**Lewis**
Acid	Donates H⊕	Accepts electron pair
Base	Accepts H⊕	Donates electron pair

$$HA \; + \; H_2O \; \rightleftharpoons \; H_3O^{\oplus} \; + \; A^{\ominus}$$

Questions:

1. (a) In the Brönsted-Lowry definition, what are the roles of an acid and a base?

   (b) Using the Brönsted-Lowry definition, which reagent in the reaction in Model 1 acts as the acid (HA or H_2O) and which acts as the base (HA or H_2O)? Explain.

2. (a) In the Lewis definition, what are the roles of an acid and a base?

   (b) Using the Lewis definition, which reagent in the reaction in Model 1 acts as the acid (HA or H_2O) and which acts the base (HA or H_2O)? Explain.

3. In the equation shown in Model 1, label the acid, base, conjugate acid and conjugate base. In this case, does it matter which definition is used to determine the acid/base roles?

4. (a) Which definition, Brönsted-Lowry or Lewis, best describes what occurs during the reaction shown below? Explain why one definition is favored for this reaction.

$$\underset{\substack{Cl \\ | \\ Cl-Al \\ | \\ Cl}}{} \quad + \quad Cl-Cl \quad \longrightarrow \quad \underset{\substack{Cl \\ | \ominus \quad \oplus \\ Cl-Al-Cl-Cl \\ | \\ Cl}}{}$$

   (b) Label each reactant as acid and base in the reaction shown in #4a above.

Model 2: Acid Dissociation Constant, Ka and pKa

The equilibrium constant (Keq) is equal to the concentration of the products over the concentration of the reactants. Since the concentration of water is constant in an acid-base reaction, the equation can be rearranged to define the acid dissociation constant, Ka:

$$Ka = [H_2O]Keq = \frac{[H_3O^{\oplus}]\,[A^{\ominus}]}{[HA]}$$ and   **pKa = -log Ka**

Questions:

5. (a)   In the equilibrium constant, label the products and the reactants. If Ka is large, the concentration of (*circle one*) products / reactants would be larger. If Ka is large, the acid is (*circle one*) strong / weak.

   (b)   Consider the definition of pKa in Model 2. If Ka is large, the pKa is (*circle one*)   small / large. If Ka is small, the pKa is (*circle one*) small / large.

   (c)   A small pKa value means the acid is (*circle one*) strong / weak and a large pKa means the acid is (*circle one*) strong / weak.

   (d)   A large pKa value means the base is (*circle one*) strong / weak.

6. Of the following pKa values, circle the one that corresponds to the strongest acid and draw a box around the one that corresponds to the strongest base.

   -10, 1, 5, 30

Model 3: pKa Values of Common Acid/Base Pairs

$$H{-}A \rightleftharpoons H^{\oplus} + A^{\ominus}$$

Acid (HA)	pKa	Conjugate Base (A⊖)
HBr	-9	Br⊖
HCl	-2.2	Cl⊖
CH₃CO₂H	4.74	CH₃CO₂⊖
H₂O	15.7	HO⊖
NH₃	33	NH₂⊖
CH₄	~50	CH₃⊖

Questions: Refer to the Lewis acid/base definition and pKa values for the following questions.

7. (a)   Which atom of acid HA in Model 3 will accept the pair of electrons? (*Circle one*) H / A.

   (b)   Of the acids listed in Model 3, which is most likely to accept an electron pair? Which acid is least likely to accept an electron pair? Explain.

   (c)   The acid most likely to accept an electron pair is the (*circle one*) stronger / weaker acid.

(d) Of the <u>bases</u> listed in the Model 3, which is <u>most</u> likely to donate an electron pair? Which base is <u>least</u> likely to donate an electron pair? Explain.

(e) The base most likely to donate an electron pair is the (*circle one*) stronger / weaker base.

8. A reaction always proceeds from stronger to weaker reagent (**SA + SB → WA + WB**). Use the pKa values in Model 3 to help answer questions about the following reaction:

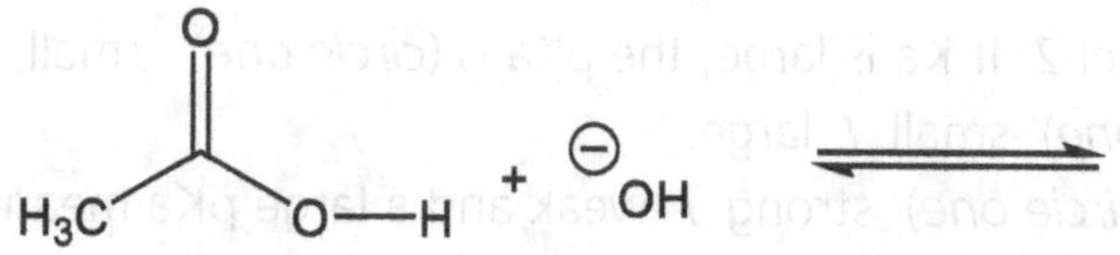

(a) Using the information from the table in Model 3, the pKa of CH_3CO_2H = _____ and the pKa of $HO^{\ominus}$ = _____. Based on these pKa values, label the acid and the base in the above reaction.

(b) Describe what happens to the electron pair that the base donates.

(c) Describe what happens to the electron pair that the acid accepts.

(d) Draw the two products formed above, and label as conjugate acid/conjugate base.

(e) <u>Group Check</u>: Does the net charge of the reactants equal the net charge of the products drawn? If not recheck your answer.

(f) Predict whether the equilibrium favors the forward or the reverse direction. Explain.

9. Weak acids (pka>6) can act as either acids or bases depending on what other reagents are present. Use the acid/base definitions and pKa values to answer the questions below.

$$H_3C-\underset{\underset{H}{|}}{O} \;+\; HO^{\ominus} \;\rightleftharpoons\; CH_3O^{\ominus} \;+\; H_2O \qquad \text{Eq. 1}$$

$$CH_3O-H \;+\; H-Br \;\rightleftharpoons\; CH_3\overset{\oplus}{O}H_2 \;+\; Br^{\ominus} \qquad \text{Eq. 2}$$

(a) Using any acid/base definition, predict whether methanol (CH_3OH) acts as the acid or the base in Eq.1. Explain.

(b) Using any acid/base definition, predict whether methanol (CH_3OH) acts as the acid or the base in Eq. 2. Explain.

(c) Label the acid, base, conjugate acid and conjugate base for each reaction.

(d) Once your group has reached consensus on the above questions, discuss and predict an **approximate** pKa value for methanol, based on pKa values listed in Model 3.

Model 4: Curved Arrows

$$\overset{:\overset{..}{O}}{\underset{H_3C\qquad CH_3}{\Big|}} \longleftrightarrow \overset{:\overset{..}{O}:^{\ominus}}{\underset{H_3C\quad\overset{\oplus}{}\quad CH_3}{\Big|}} \qquad \text{Eq. 3}$$

$$H-\overset{..}{\underset{..}{Br}}: \longrightarrow H^{\oplus} \;+\; :\overset{..}{\underset{..}{Br}}:^{\ominus} \qquad \text{Eq. 4}$$

$$H^{\oplus} \;+\; H-\overset{..}{\underset{..}{O}}-H \longrightarrow H-\overset{\overset{\displaystyle H}{|}}{\underset{..}{O}}{}^{\oplus}-H \qquad \text{Eq. 5}$$

Questions:

10. (a) Consider the resonance structures shown in Eq. 3. Describe what the curved arrow shows. (Be specific)

(b) Consider the dissociation of HBr shown in Eq. 4. Describe what the curved arrow shows. (Be specific)

(c)   Consider the acid/base reaction shown in Eq. 5. Describe what the curved arrow shows. (Be specific)

(d).   Curved arrows show movement of (*circle one*) protons / electrons / atoms.

11. Once your group has reached consensus on the above questions, come up with a description of how to draw curved arrows (i.e. where to start the arrow and where to end it).

12. Draw curved arrows to show formation of the products from questions #8 and #9 above

13. Use of curved arrows is an important tool in showing how organic reactions proceed. Based on your answer to #11, explain why the Lewis acid/base definition makes more sense.

Reflection: on a separate sheet of paper.
As a group, describe three concepts your group has learned from this activity and the one most important unanswered question about this activity that remains with your group. Turn this in before leaving class.

Additional Questions:
14. Draw the products of the reaction between HBr and H_2O in the space below.
    (a)   Label the acid, base, conjugate acid and conjugate base. (Use the Lewis acid/base definition).

    (b)   Draw in all lone pairs, then use curved arrows to show the transfer of electrons.

15. Identify the acid and the base in each reaction, draw all the products expected and draw curved arrows to show electron movement. Predict if the reaction will go in the forward or the reverse direction. Unless otherwise given, pKa values can be estimated from values in Model 3.

    (a)

(b)

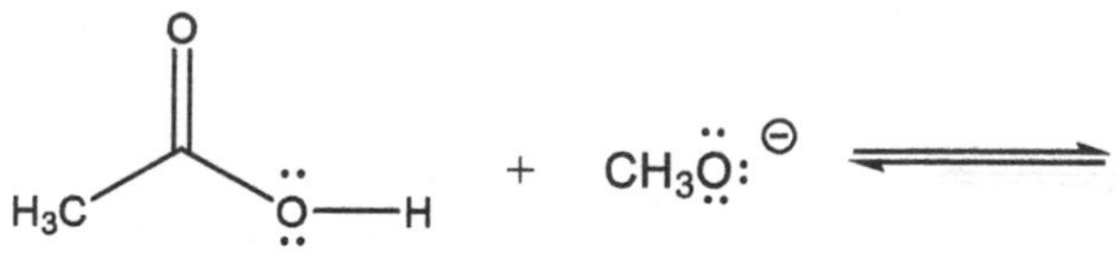

$$H_3C-C(=O)-\overset{..}{\underset{..}{O}}-H \quad + \quad CH_3\overset{..}{\underset{..}{O}}:^{\ominus} \quad \rightleftharpoons$$

(c)

$$H_2\overset{..}{N}:^{\ominus} \quad + \quad H-C\equiv C-H \quad \rightleftharpoons$$

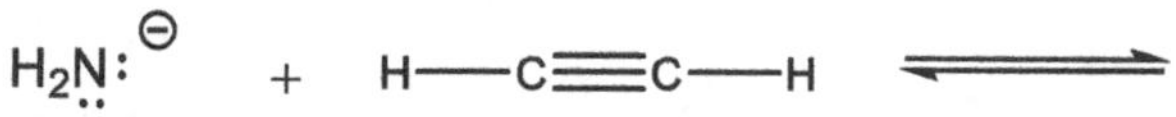

pKa = 25

(d)

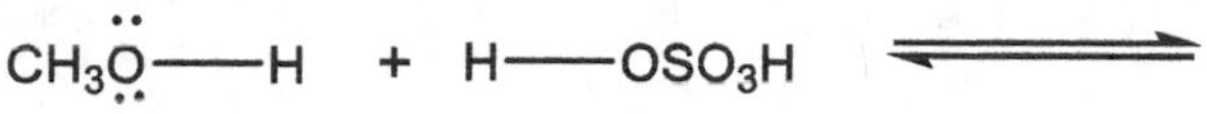

$$CH_3\overset{..}{\underset{..}{O}}-H \quad + \quad H-OSO_3H \quad \rightleftharpoons$$

(e)

$$+ \ Na\overset{..}{\underset{..}{O}}H \quad \rightleftharpoons$$

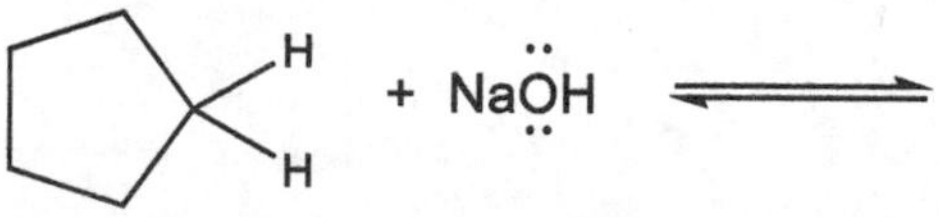

(f) Aspirin (acetylsalicylic acid), isolated from the bark of the willow tree, is used to relieve minor aches and pains, reduce fevers, and reduce swelling.

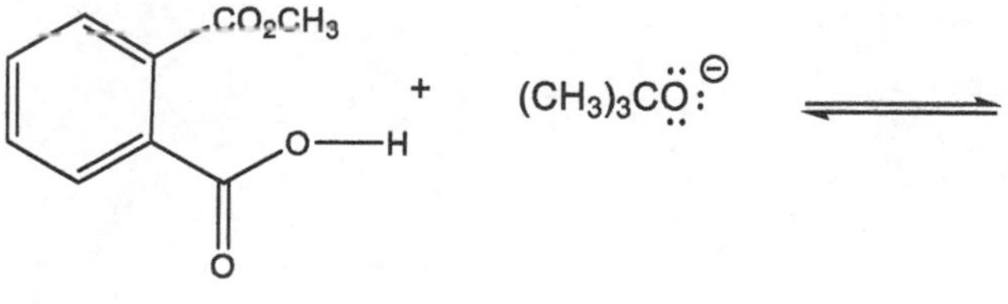

$$+ \quad (CH_3)_3C\overset{..}{\underset{..}{O}}:^{\ominus} \quad \rightleftharpoons$$

(g) Gallic acid, an organic acid, has anti-fungal, anti-viral and anti-oxidant properties.

Class Activity 3B

Acids and Bases
Part B: Predicting Acid/Base Strength Without pKa Values

Prior Knowledge:
Before beginning this activity, students should be familiar with the following concepts:

Learning Objectives
Content Learning Objectives:
After completing this activity, students should be able to:
- Predict the strength of an acid based on the stability of the conjugate base.
- Determine how factors such as electronegativity, formal charge, resonance, size and induction affect the stability of the conjugate base.

Process Objectives:
- Information Processing. Students interpret and manipulate factors (size, formal charge, electronegativity, resonance, and induction) to predict the stability of a conjugate base.
- Critical Thinking: Students evaluate the conjugate base stabilities in order to conclude which acid is stronger when comparing several different acids.

Class Activity 3B

Acids and Bases
Part B: Predicting Acid/Base Strength Without pKa Values

Model 1: Stability of the Conjugate Base
The strength of an acid can be predicted by estimating the stability of the conjugate base formed. The stability of a base is affected by the following factors: **size, formal charge, electronegativity, resonance stabilization and inductive electron withdrawing effects.**

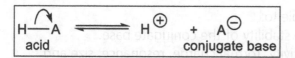

Questions: (Refer to the Lewis acid/base definition for the following questions.)

1. Consider the <u>forward</u> direction of the reaction in Model 1:

   (a)   Based on the Lewis definition, acid HA (*circle one*) donates / accepts an electron pair.

   (b)   If the conjugate base A⊖ is stable, HA is (*circle one*) more / less likely to accept an electron pair.

   (c)   If the conjugate base A⊖ is unstable, HA is (*circle one*) more / less likely to accept an electron pair.

   (d)   The more stable the conjugate base, then the acid it is derived from is (*circle one*) stronger / weaker.

2. (a)   Draw the conjugate base produced on dissociation of the acids in the following two reactions. Calculate the formal charge of oxygen for each conjugate base.

   (b)   Which conjugate base above would you expect to be more stable and why?

   (c)   Based on the stability of the conjugate base, which acid would be the strongest, $H_3O^{\oplus}$ or H_2O? Explain.

3. Consider the <u>reverse</u> direction of the reaction in Model 1:

(a)   If A$\ominus$ is stable, it is (*circle one*) more / less likely to donate an electron pair to form HA.

(b)   If A$\ominus$ is unstable, it is (*circle one*) more / less likely to donate an electron pair to form HA.

(c)   An unstable base is (*circle one*) more / less reactive.

(d)   Based on the stability of the conjugate base, predict the direction of the equilibrium for each reaction shown in #2a above.

4. (a)   Draw the conjugate base produced on dissociation of CH_4 shown below. Calculate the formal charge of carbon in the conjugate base.

H—C—H with H above and H below, double arrow (equilibrium) pointing right

(b)   Which would you expect to be more stable, CH_4 or the conjugate base? Why?

(c)   Predict the direction of the equilibrium based on your answer above.

(d)   The pKa for methane (CH_4) is approximately 50. Does this agree with your conclusions on the stability of the conjugate base? Explain.

5. Once your group has reached consensus on the above questions, discuss how the stability of a base affects the strength/reactivity of a conjugate base and the strength/reactivity of the acid it is derived from.

Model 2: Acid/Base Strength and Size

Element	pKa	HA → H⊕ + A⊠
F	3.2	HF → H⊕ + F⊠
Cl	-2.2	HCl → H⊕ + Cl⊠
Br	-9	HBr → H⊕ + Br⊠
I	-10	HI → H⊕ + I⊠

Larger

Questions:

6. From the information shown in Model 2:
   (a)   Which element is the largest in size? _____ Which element is the smallest? _____

   (b)   How does the size of the element relate to the pka values given?

   (c)   Based on the pKa values above, which acid is stronger, HF or HI? The electronegativity of F = 4.0 while the electronegativity of I = 2.7. Based on the pKa values, which effect is stronger in determining acid strength, electronegativity or size? Explain your answer.

   (d)   As a group summarize your conclusions regarding how size affects the stability of the base. (Hint: use the term "polarizability" in your answer.)

   (e)   Compare the two conjugate bases shown below. Which base would be expected to more stabilized by size? (tBuO⊠ or Br⊠). Which acid would be a stronger acid, tBuOH or HBr? Explain your answer.

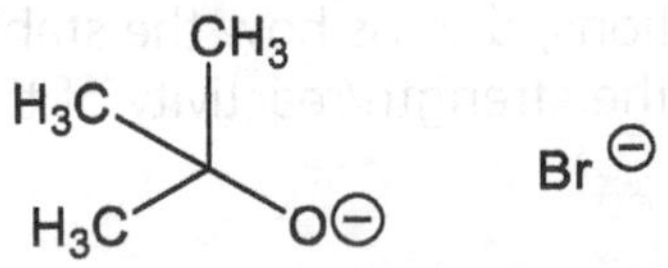

H_3C — C(CH_3) — O⊖   Br ⊖

Model 3: Acid/Base Strength and Electronegativity Values

Element	Electronegativity	pKa	HA → H⊕ + A⊠
F	4.0	3.2	HF → H⊕ + F⊠
O	3.4	15.7	HOH → H⊕ + HO⊠
N	3.0	36	HNH₂ → H⊕ + H₂N⊠
C	2.5	40	HCH₃ → H⊕ + H₃C⊠

Compare electronegativity of elements of similar size.

Questions:

7. (a) The elements shown in Model 3 are on the same row of the periodic table. Would the size of the elements be approximately the same? (*circle one*) yes / no

(b) Based on electronegativity values, which element in Model 3 is the most electronegative? Which is least electronegative?

(c) An electronegative element (*circle one*) attracts / repels electrons.

(d) In the general reaction **HA → H⊕ + A⊠**, which conjugate base (A⊠) would be more stable, the one derived from the (*circle one*) most / least electronegative element?

(d) Rank the conjugate bases (F⊠, HO⊠, H₂N⊠, H₃C⊠) from most stable to least stable.

(e) Once your group agrees on the above questions, write a statement that describes how electronegativity affects the stability and the strength of the conjugate base.

Model 4: Acid/Base Strength and Inductive Electron Withdrawing Effects

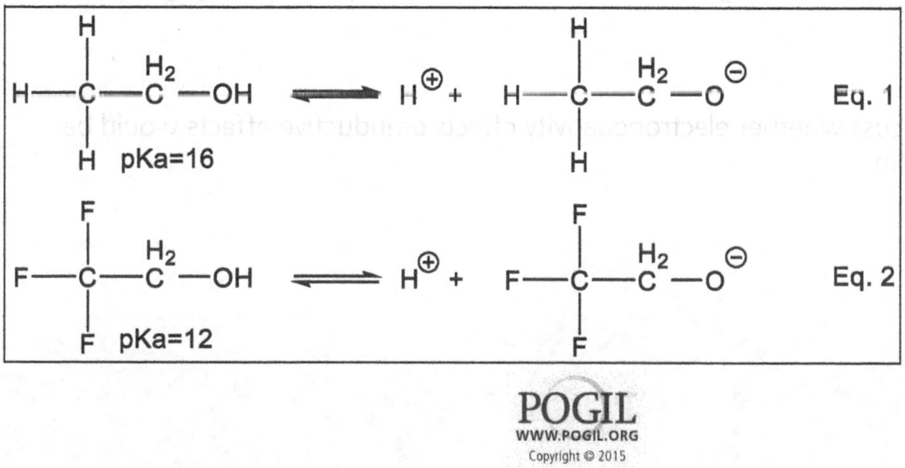

POGIL
WWW.POGIL.ORG
Copyright © 2015

Questions:

8. (a)   Based on the pKa values in Model 4, which alcohol is the stronger acid, ethanol in Eq. 1, or trifluoroethanol in Eq. 2?

    (b)   Compare the conjugate bases in Eq. 1 and Eq. 2. What is different? What is the same?

    (c)   Consider the conjugate base in Eq. 2. The electrons in the covalent C—F bond would be (*circle one*) equally / unequally shared.

    (d)   Label the C—F bond with $\delta\ominus$ and $\delta\oplus$ to indicate partial charges. What effect would a polar bond like C—F have on a nearby negative charge?

    (e)   Explain how the presence of the CF_3 group in Eq. 2 helps stabilize the conjugate base.

    (f)   Predict whether <u>difluoroethanol</u> would be more or less acidic than <u>trifluoroethanol</u>. Make sure that everyone in your group agrees with this conclusion before moving on to the next question.

9. A difference in electronegativity between atoms causes an <u>inductive electron withdrawing effect</u>, shown by a dipole arrow.

    (a)   Draw dipole arrows for each of the polar bonds in the following bases.

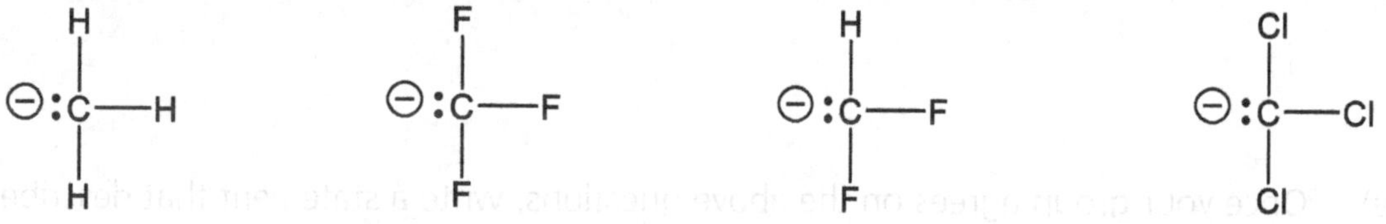

    (b)   Rank the above bases from most stable (1) to least stable (4) due to inductive effects.

    (c)   Draw the acid that each of the above bases would be derived from. Predict which acid would be more acidic (strongest acid) based on the stability of the conjugate base. Explain.

    (d)   As a group discuss whether electronegativity effects or inductive effects would be stronger. Explain.

Model 5: Acid/Base Strength and Resonance

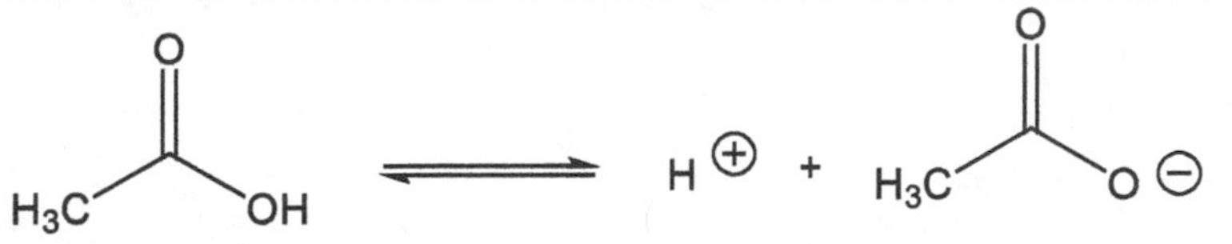

acetic acid

Eq. 3

$$CH_3CH_2{-}OH \rightleftharpoons H^{\oplus} + CH_3CH_2O^{\ominus}$$

ethanol

Eq. 4

Questions:

10. (a) Identify the conjugate base for each of the reactions shown in Model 5.

(b) Compare the conjugate base from each reaction. Focus on the atom that contains the charge and determine if there are any differences in electronegativity, charge or size.

(c) Draw all possible resonance structures for each conjugate base shown in Model 5.

(d) Resonance forms show that electrons can be spread out or delocalized, which has a stabilizing effect. What conclusions can be made about how resonance stabilization affects stability of the conjugate base?

(e) Based on your answer for #10d, which would be more acidic, acetic acid or ethanol? Explain.

11. (a) Draw the conjugate base for each of the following acids (by removing the hydrogen indicated with the arrow). Draw all possible resonance forms for each conjugate base.

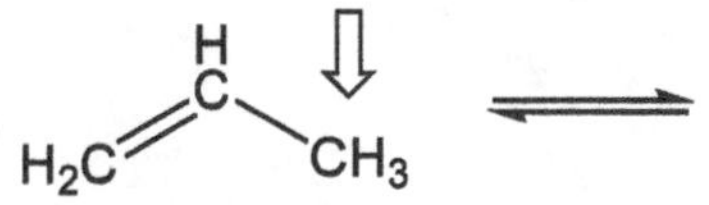

A

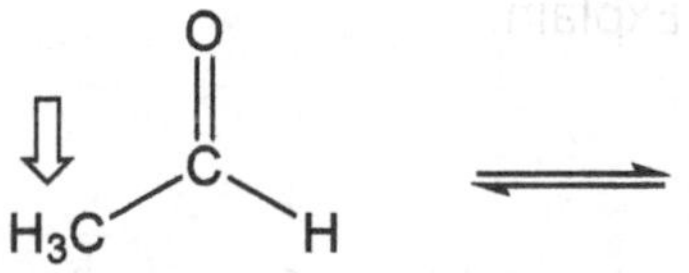

B

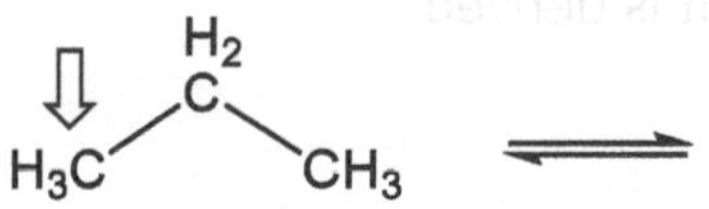

C

(b) Predict which base would be the most stable and which would be the least stable based on resonance stabilization. (Hint: in general the more resonance structures, the more stable.)

(c) Once your group has reached agreement on the above questions, predict which acid would be the most acidic and which would be the least acidic. Explain.

Reflection: on a separate sheet of paper.
As a group, describe three concepts your group has learned from this activity and the one most important unanswered question about this activity that remains with your group. Turn this in before leaving class.

Additional Questions:

12. (a) Draw the conjugate base for each of the following acids.

(b) Predict which conjugate base would be more stable. Explain.

(c) Predict which acid would be more acidic. Explain how the stability of the conjugate base helps you predict the acidity of the acid from which it is derived

13. For the following compound:

(a)   Draw the conjugate base expected when one Ha is removed.

(b)   Draw the conjugate base expected when one Hb is removed.

(c)   Draw the conjugate base expected when Hc is removed.

(d)   Determine which proton is more acidic, Ha, Hb or Hc. Explain your reasoning.

14. Rank the following compounds from most acidic to least acidic. Most=1, least = 3
    (a)   ClCH₂CH₂OH, CH₃CH₂OH, CH₃CH₂NH₂

    (b)   CH₃CH₂CH₃, CH₃CH₂OH, CH₃CH₂CO₂H

    (c)   BrCH₂CO₂H, CH₃CH₂CO₂H, CH₃CH₂CH₂OH

    (d)   CH₃CH₂NH₂, (CH₃)₃N, CH₃CH₂OH

15. Rank the following compounds from most basic to least basic. Most basic (least stable)=1, least basic (most stable)=3
    (a) $CH_3CH_2\Theta$, $CH_3O\Theta$, $CH_3NH\Theta$

    (b) $CH_3\Theta$, $HO\Theta$, $Br\Theta$

    (c) $CH_3CO_2\Theta$, $CH_3CH_2O\Theta$, $BrCH_2CO_2\Theta$

16. (a)   Predict which proton in ascorbic acid (vitamin C) would be the most acidic. Explain using the stability of the conjugate base produced by removing each proton.

    (b)   Explain why ascorbic acid is slightly more acidic than acetic acid (CH_3CO_2H).

17. Predict which analgesic, acetaminophen or aspirin is more acidic. Explain.

Acetaminophen (Tylenol)   Acetylsalicylic acid (Aspirin)

Class Activity 4

Nomenclature of Alkanes

Prior Knowledge:
Before beginning this activity, students should be familiar with the following concepts:

- Drawing condensed and line-angle organic structures

Learning Objectives
Content Learning Objectives:
After completing this activity students should be able to:
- Identify the parent chain in cyclic and acyclic alkanes.
- Determine the IUPAC name of alkane compounds.
- Produce both common and IUPAC names for more complicated substituents.

Process Objectives:
- Problem solving. Students execute a strategy in order to determine the names of organic compounds.

Class Activity 4

Nomenclature of Alkanes

Model 1: Straight Chain Alkanes

# Carbons	Prefix	Name	Structure
1	Meth	Methane	CH_4
2	Eth	Ethane	CH_3CH_3
3	Prop	Propane	$CH_3CH_2CH_3$
4	But	Butane	$CH_3CH_2CH_2CH_3$
5	Pent	Pentane	$CH_3CH_2CH_2CH_2CH_3$
6	Hex	Hexane	$CH_3CH_2CH_2CH_2CH_2CH_3$
7	Hept	Heptane	$CH_3CH_2CH_2CH_2CH_2CH_2CH_3$
8	Oct		$CH_3CH_2CH_2CH_2CH_2CH_2CH_2CH_3$
9	Non		$CH_3CH_2CH_2CH_2CH_2CH_2CH_2CH_2CH_3$
10	Dec		$CH_3CH_2CH_2CH_2CH_2CH_2CH_2CH_2CH_2CH_3$

Questions:

1. Based on the information in Model 1:
   (a) What do all the names have in common?

   (b) What is unique about each of the compound names?

   (c) What information does the prefix of a name provide?

   (d) In general, how is the prefix modified to name a straight chain alkane?

   (e) Write in the names in the table for compounds with 8, 9 and 10 carbons.

2. The prefix <u>dodec</u> means twelve carbons. Give the name of the corresponding alkane and draw the structure.

Model 2: Branched Alkanes

	Parent Chain	#C in Parent	Structure	Substituent	#C in Substituent	Compound Name
A	propane	3		methyl	1	2-methylpropane
B						2-methylbutane
C	pentane	5		ethyl	2	
D						3-methylhexane

Questions:

3. For entry A in Model 2.
    (a) How many carbons are in the parent chain?　　　Circle the parent chain in the structure shown. Underline the parent in the compound name.
    (b) How many carbons are in the substituent?　Put a box around the substituent in the structure shown.
    (c) Number the carbons in the parent chain consecutively.
    (d) Why is there a number 2 in front of the methyl in the name?

    (e) For 2-methylpropane, what punctuation mark separates a number from a letter?

    (f) In your group discuss what factors distinguish the parent from the substituent.

4. Complete the table for B, C and D.

5. (a) For compound B in Model 2, circle and number the carbons in the parent chain.
    (b) Does your numbering scheme agree with the name given?
    (c) If not, re-number the chain so the numbering agrees with the name.

(d)   The name of compound B is 2-methylbutane, not 3-methylbutane. What conclusion can be made for numbering substituents?

6. (a)   For compound D, circle the parent chain. Does the parent have six carbons?
    (b)   If not, circle a chain with six continuous carbons.
    (c)   The parent name for compound D is 3-methylhexane not 2-ethylpentane. What conclusion can be made about choosing the parent chain?

7. Once your group agrees on the contents of Model 2, devise a set of rules for naming substituents using the prefixes found in Model 1.

Model 3: Multiple Substituents

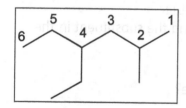

Questions:

8. (a)   For the compound in Model 3, what is the name of the parent chain?
    (b)   What are the names of the substituents?

    (c)   If there is more than one substituent in a compound, substituent names are listed in alphabetical order. Provide the name of the compound above (remember to include the numbers for the substituents).

    (d)   Once your group has reached agreement on a name for the compound in Model 3, check the name to make sure it has followed the naming rules listed below.
          √ longest chain for parent
          √ substituents have lowest possible numbers
          √ substituents in alphabetical order

(e)   Draw the structure of 3-ethyl-2-methyl-4-propyloctane.

Model 4: Multiple Identical Substituents

2,3-dimethylbutane

A   B

Questions:

9. For compound A in Model 4:

(a)   How many carbons are in the parent chain?   Circle the parent on structure A.

(b)   How many substituents are there in compound A? _____
Are the substituents identical? (Circle one) yes / no.
In the compound name, what prefix is used to indicate how many of the same substituent there are in the compound? _____

(c)   Number the parent carbons; what are the substituent numbers? ___

(d)   What punctuation is used to separate numbers in a name?

10. For compound B in Model 4:

(a)   How many carbons are in the parent chain?   Circle the parent on structure B.

(b)   Number the parent from right to left, what are the substituent numbers?

(c)   Number the parent from left to right, what are the substituent numbers?

(d)   Which numbering scheme (from #10b or #10c) do you think is correct? Explain.

(e)   Provide the correct name for compound B.

(f)   Draw the structure for 3-ethyl-2,4-dimethyloctane

Model 5: Branched Substituents

Branched substituents are named using IUPAC or common names. For simplicity, the parent compounds below are not specified, but drawn as rectangles. Only the substituent is named, and these names would precede the parent name.

Substituent	IUPAC Name in ( )	Common Name	Skeletal
CH₃ ²CH₃——¹CH [parent]	(sub-parent underlined) (1-methyl<u>ethyl</u>)	isopropyl	
CH₃ ²CH₃——¹C——CH₃ [parent]	(1,1-dimethyl<u>ethyl</u>)	tert-butyl	
³H₃C CH₃ ²CH ¹CH₂ [parent]	(2-methyl<u>propyl</u>)	isobutyl	
H₃C³ ²CH₂ HC——CH₃ ¹ [parent]	(1-methyl<u>propyl</u>)	sec-butyl	

Questions:

11. (a)   For each of the examples shown in Model 5, look at the bond between the parent compound and the substituent. What number is always given to the carbon of the substituent that is attached to the parent? _____

   (b)   <u>Starting</u> with this point of attachment to the parent chain, find the longest continuous chain in the substituent and circle it (for all compounds in Model 5).

   (c)   Since the portion you have circled belongs to the <u>substituent</u>, we will call it the "sub-parent". Does the sub-parent circled agree with the sub-parent name underlined in the column labeled IUPAC name? If not, make corrections to the circled sub-parent in part (b) above.

   (d)   Once the sub-parent is found, any substituent names are placed (*circle one*) before / after, the sub-parent. Make sure your group agrees with the IUPAC names of each substituent before moving on.

   (e)   Look at the common names given for each substituent. What does the sub-parent name refer to (i.e. for isopropyl, why is propyl chosen as the sub-parent)?

   (f)   To the right of the table, draw each substituent in the skeletal form. Make any notes about the shape of this substituent that might help you remember the common name.

12. Provide <u>two</u> names for the following compound using IUPAC substituent names for one and common substituent names for the other.

Model 6: Cyclic Alkanes
Cyclic alkanes are named by adding the prefix cyclo to the alkane base.

Questions:

13. (a)   How many carbons are in the <u>ring</u> in the above structure?

(b)   How many carbons are in each group?

(c)   Which has more carbons, the ring, group 1 or group 2? The one having the most number of carbons is the parent chain.

(d)   Name the parent chain for the above compound. (Remember to use cyclo if the ring is the parent).

(e)   Name the two substituents (all substituents end in -yl).

14. (a)   In rings, one substituent will always be at carbon 1. Number the carbons in the ring 1-6 starting with group 1 and continue so that group 2 has the lowest number possible. (NOTE: you are choosing between clockwise or counterclockwise).

(b)   Number the carbons in the ring 1-6 starting with group 2 and continue so that group 1 has the lowest number possible. (NOTE: you are choosing between clockwise or counterclockwise).

(c)   Do either of the two numbering schemes from (a) or (b) provide lower position numbers for the substituents?

(d)   If both numbering schemes give the same numbers, chose the scheme that gives the lower number to the substituent that occurs earlier alphabetically. Once your group agrees on the above questions, provide the complete name of the compound shown in Model 6.

15. For the following compound:

(a) Determine if the parent name is that of the ring or that of the group attached to the ring. Explain.

(b) Circle the parent compound.

(c) Number the parent so that the substituents have the lowest possible numbers.

(d) Provide an acceptable name for the compound.

Reflection: on a separate sheet of paper.
As a group, describe three concepts your group has learned from this activity and the one most important unanswered question about this activity that remains with your group. Turn this in before leaving class.

Additional Questions:
16. The following two structures have been named incorrectly. Give the correct name for each compound and explain using the rules developed above.

Structure	Incorrect name	Correct name
	2-ethylhexane	
	4-methylpentane	

17. Using the rules devised above, provide names for each of the following.

18. Draw structures for the following compounds:
    (a) 3,4-dimethylhexane

    (b) 5-tert-butyl-7-ethyl-2,3-dimethylnonane

    (c) 3-ethyl-4-isobutyl-2,7-dimethyldecane

    (d) 2-ethyl-1,1-dimethylcyclopentane

Class Activity 5A

Conformations of Alkanes
Part A: Acyclic Compounds

Prior Knowledge:
Before beginning this activity, students should be familiar with the following concepts:

- Drawing Lewis, condensed and line-angle organic structures.
- Definition of isomers (from general chemistry).
- Potential energy diagrams.

Learning Objectives
Content Learning Objectives:
After completing this activity students should be able to:
- Describe the difference between constitutional and conformational isomers.
- Draw conformations in wedge dash, sawhorse and Newman projections and determine the difference between staggered and eclipsed.
- Predict the relative energy of conformations from comparing conformations in Newman projections.

Process Objectives:
- Information Processing. Students interpret, manipulate and transform wedge-dash, sawhorse and Newman representations of organic compounds.
- Critical Thinking. Students evaluate different representations of conformational isomers to arrive at conclusions regarding stability of various conformations.

Class Activity 5A

Conformations of Alkanes
Part A: Acyclic Compounds

Model 1: Isomers

Constitutional Isomers: atoms are connected differently.
Conformational Isomers: atoms are connected the same, but differ in the rotation around single bonds.

Questions:

1. (a)   For the structures in Model 1, write down the order of atom connectivity for compounds B and C (ignore the hydrogens). Compound A is given as an example.

   A: C—C—C—O

   B:

   C:

   (b)   Is the order of connectivity the same for structures A and B? *(Circle one)* yes / no.

   (c)   Is the order of connectivity the same for structures B and C? *(Circle one)* yes / no.

   (d)   Is the order of connectivity the same for structures A and C? *(Circle one)* yes / no.

   (e)   Using the definition of constitutional isomers, indicate which of the pairs from Model 1 are constitutional isomers.

   (Circle all that apply)   A and B   B and C   A and C

2. (a)   In Model 1, what does a wedge represent? What does a dash represent?

   (b)   Compare structures A and C. Are the atoms connected in the same order?

   (c)   What is different between structures A and C?

   (d)   Based on the definitions in Model 1, would A and C be constitutional or conformational isomers? Make sure everyone in your group agrees before moving on.

Model 2: Conformations of Ethane

staggered

wedge/dash   sawhorse   Newman

eclipsed

Questions:

3. In the wedge dash representation of ethane, how are the hydrogens drawn to show their spatial orientation?

4. Envision the wedge dash model where the C-C bond is in the plane of your paper (use a pen to simulate the bond). Visually rotate the whole molecule by 90°, so that the C-C bond is now perpendicular to the paper. This is the sawhorse representation. On the sawhorse model above, write an "F" near the <u>front</u> carbon and a "B" near the <u>back</u> carbon. Circle the bond that was rotated by 90°.

5. The Newman projection is the same as a sawhorse projection that has just been compressed along the C-C bond. How is the carbon in the <u>front</u> represented in a Newman projection? How is the carbon in the <u>back</u> represented?

6. (a)   From Model 2, compare the two conformations of ethane, <u>staggered</u> and <u>eclipsed</u>. What is different between these conformations?

   (b)   Once your group agrees on the above questions, discuss which conformation would be the most stable. Explain your reasoning.

   (c)   Describe why the names staggered and eclipsed represent the forms shown.

Model 3: Newman Projections

Newman projections clearly show the dihedral angle, θ, which defines the difference between the staggered and eclipsed conformations.

$\theta=$ ___   Ha   Hb $\theta=$ ___   Ha   Hb $\theta= 0°$ Hc

staggered
gauche $\theta= 60°$
anti $\theta= 180°$

eclipsed
$\theta= 0°$

dihedral angle, θ

Questions:

7. Using the information in Model 3, identify the dihedral angle for the following:

   Staggered gauche $\theta =$

   Staggered anti $\theta =$

   Eclipsed $\theta =$

8. For the staggered form in Model 3:

   (a) What is the dihedral angle between Ha and Hb (60° or 180°)? Is this gauche or anti?

   (b) What is the dihedral angle between Ha and Hc (60° or 180°)? Is this gauche or anti?

9. (a) A Newman projection is draw by looking straight down a particular C—C bond. The carbon furthest away is drawn as a large circle, while the carbon closest is represented as a small circle or the center of the Y shape. Draw a Newman projection of <u>propane</u> as viewed along the C1-C2 bond. (HINT: draw C1 in front, C2 in back, then add all substituents).

   (b) Once your group agrees on the Newman projection drawing of propane, draw both the staggered and eclipsed forms of the Newman projection above. Circle the form that is most stable.

Model 4: Potential Energy Diagram for Conformations of Ethane

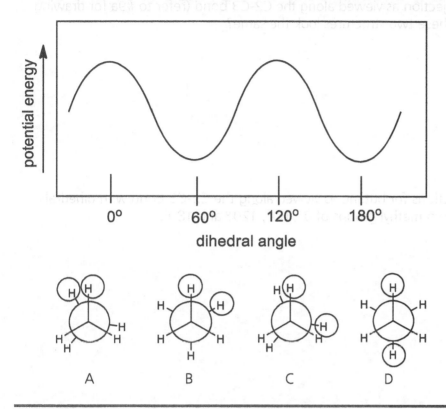

Questions:

10. Consider the energy diagram shown in Model 4.

    (a) What is the dihedral angle between the circled hydrogens in projection A?

         What is the dihedral angle between the circled hydrogens in projection B?

         When converting structure A to structure B: which carbon was rotated? (*Circle one*) front / back.

         As you rotate from A → B, then B → C, what happens to the conformations? (HINT: use terms from Model 3.)

    (b) Do the Newman projections with a dihedral angle of 0° and 120° have the same energy? (*Circle one*) yes / no.
         If all hydrogens are equivalent, would you be able to distinguish between these two forms? (*Circle one*) yes / no.

    (c) Do the Newman projections with a dihedral angle of 120° and 180° have the same energy? (*Circle one*) yes / no.
         If all hydrogens are equivalent, would you be able to distinguish between these two forms? (*Circle one*) yes / no.

11. Draw a staggered Newman projection for butane (four carbons) as viewed along the C1-C2 bond AND a staggered projection as viewed along the C2-C3 bond (refer to #9a for drawing Newman projections). Do these two structures look the same?

12. (a)   Draw Newman projections for butane as viewed along the C2-C3 bond with dihedral angles between the two methyl groups of 0°, 60°, 120° and 180°.

    (b)   Identify which of the above Newman projections are staggered and which are eclipsed.

    (c)   Once your group has reached agreement on the Newman projections, discuss whether you would expect all the staggered forms to have the same energy. Explain.

    (d)   Would you expect all the eclipsed forms to have the same energy? Explain.

    (e)   Which Newman projection would be the most stable, if any? Explain.

Reflection: on a separate sheet of paper.
As a group, describe three concepts your group has learned from this activity and the one most important unanswered question about this activity that remains with your group. Turn this in before leaving class.

Additional Questions:

13. Draw a Newman projection of the most stable conformational isomer of 2, 4-dimethylhexane as viewed along the C3-C4 bond.

14. Convert the following compound into a Newman projection as viewed along the C2-C3 bond. Draw the most stable conformation.

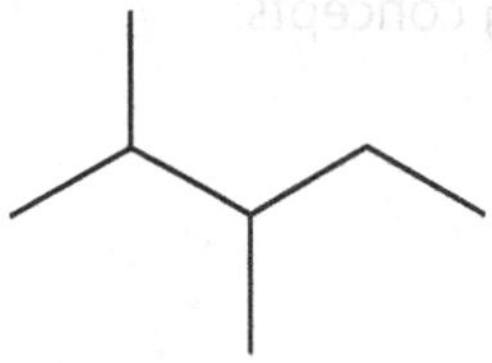

15. Convert the following Newman projections into line-angle formulas.

(a)

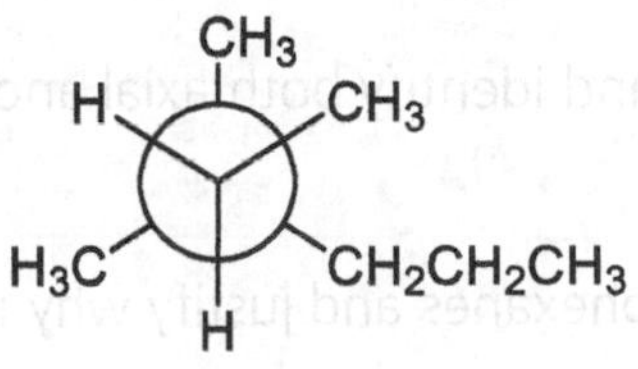

(b)

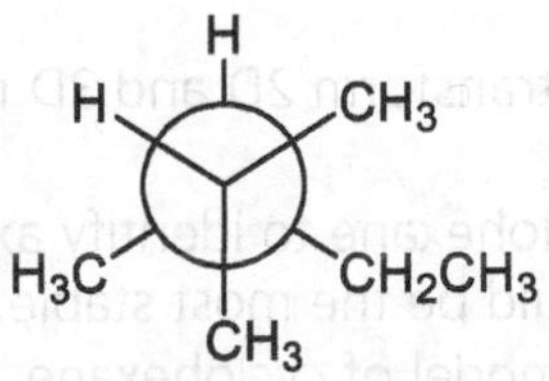

(c)

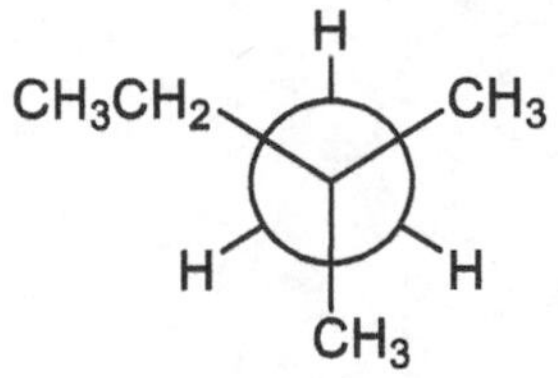

Class Activity 5B

Conformations of Alkanes
Part B: Cyclohexane Compounds

Prior Knowledge:
Before beginning this activity, students should be familiar with the following concepts:

- Drawing Lewis, condensed and line-angle organic structures.
- Constitutional and conformational isomers.
- Wedge-dash, sawhorse and Newman projections.
- Staggered and eclipsed conformations.
- Potential energy diagrams.

Learning Objectives
Content Learning Objectives:
After completing this activity students should be able to:
- Draw the boat and chair conformations of cyclohexane and identify both axial and equitorial positions.
- Identify cis and trans cyclohexane isomers.
- Predict the most stable conformation of substituted cyclohexanes and justify why it is more stable.

Process Objectives:
- Information Processing. Students interpret, manipulate and transform 2D and 3D models of cyclohexane.
- Critical Thinking. Students analyze the conformations of cyclohexane to identify axial and equatorial positions and determine which conformation would be the most stable.
- Teamwork. Students interact with each other to build one model of cyclohexane.

Class Activity 5B

Conformations of Alkanes
Part B: Cyclohexane Compounds

Model 1: Conformations of Cyclohexane

Cyclohexane has a non-planar conformation with bond angles close to the preferred angle of 109.5°. The two main conformations of cyclohexane are shown below.

(note that the dotted line is <u>not</u> a bond)

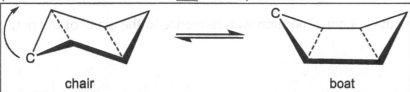

chair   boat

Questions:

1. What are the names of the two main conformations of cyclohexane?

2. The chair conformation can be envisioned as being two triangles that are connected by two parallel lines, as shown above. Explain how the arrow shows conversion from the chair to the boat conformation.

3. Using a model kit, build one model of cyclohexane in the chair conformation and a second model in the boat conformation. <u>Check with an instructor</u> to make sure your model is correct.

4. Rotate the left side of the chair in your model as indicated in Model 1 above to form the boat conformation. Compare to the model of the boat that you made. Are they the same?

5. Once your group reaches consensus on the above questions, predict whether the boat or the chair conformation is the most stable. Explain.

Model 2: Axial and Equatorial

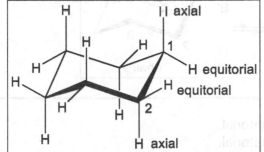

Questions:
6. In Model 2, focus on the labeled hydrogens on carbons 1 and 2.
   (a) How many axial hydrogens are on carbon 1? _____ How many equatorial? _____
   (b) How many axial hydrogens are on carbon 2? _____ How many equatorial? _____

   (c) Based on your answers to (a) and (b) what can you conclude about the number of axial and equatorial positions on each carbon?

   (d) Describe how to locate the axial position with reference to the plane of the ring.

   (e) Describe how to locate the equitorial position with reference to the plane of the ring.

   (f) Label each of the remaining hydrogens in the cyclohexane above as axial and equatorial.

7. On your model of cyclohexane, identify the axial and equatorial hydrogens.

8. Find any axial hydrogen that is pointing up. On the adjacent carbon, which direction is the axial hydrogen pointing?

9. Find any equatorial hydrogen that is pointing upwards. On the adjacent carbon, which direction is the equatorial hydrogen pointing?

10. In your group discuss what conclusions can be made regarding the axial and equatorial hydrogens on adjacent carbons?

Model 3: Chair-chair Interconversion or Ring Flip
Rotation around the single bonds of cyclohexane results in conversion between chair conformations.

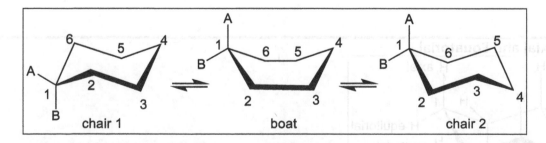

| chair 1 | boat | chair 2 |

Questions:
11. In chair 1 above, group A is (*circle one*) axial / equatorial.
12. In chair 2 above, group A is (*circle one*) axial / equatorial.

13. What conclusion can be made regarding the orientation of a group when a ring flip occurs?

14. Consider group B in both chair 1 and chair 2. Does the conclusion you derived in question #13 agree with the orientation of group B when a ring flip occurs?

15. Once your group reaches consensus on the above questions, predict whether there would be any energy differences between chair 1 and chair 2 if A and B are both hydrogen. Explain.

16. On the left below is one chair conformation of methylcyclohexane.

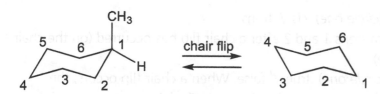

(a)   Identify whether the methyl group is in the axial or the equatorial position.

(b)   Complete the drawing on the right, showing the structure that is formed after a chair flip has occurred.

(c)   Identify whether the methyl group in the structure on the right is in the axial or the equatorial position.

17. Of the two chair forms of methylcyclohexane, construct an explanation for why the conformation with the methyl group in the equatorial position is more stable than the conformation with methyl group axial.

Model 4: Disubstituted Cyclohexanes

Recall that the spatial relationship between two substituents on different carbons in a ring must be identified as either cis (same) or trans (opposite).

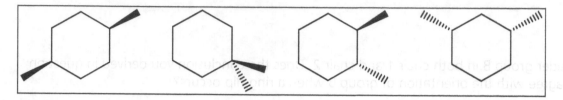

Questions:

18. Label each of the dimethylcyclohexanes shown in Model 4 as cis, trans or neither.

19. Consider the chair conformation of 1, 2-dimethylcyclohexane on the left below.

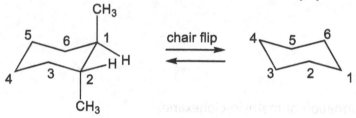

(a)   The two methyl groups are (*circle one*) cis / trans.

(b)   Draw the methyl groups at carbons 1 and 2 after a chair flip has occurred (on the chair form on the right side above).

(c)   The following statement is (*circle one*) true / false. When a chair flip occurs, the relationship between two substituents remains the same. Explain your answer.

20. Draw the <u>two</u> chair conformations for *trans*-1, 4-dimethylcyclohexane. Which conformation would you expect to be the most stable? Explain.

Reflection: on a separate sheet of paper.

   As a group, describe three concepts your group has learned from this activity and the one most important unanswered question about this activity that remains with your group. Turn this in before leaving class.

Additional Questions:

21. Draw the two chair conformations for cis-1, 4-dimethylcyclohexane. Which conformation would you expect to be the most stable? Explain.

22. If two groups in a disubstituted cyclohexane are not the same, then the more stable conformation is the one in which the <u>largest</u> group is equatorial. Draw the most stable chair conformation for each of the following compounds.

    (a)   cis-1-tert-butyl-4-methylcyclohexane

    (b)   trans-1-tert-butyl-4-methylcyclohexane

    (c)   trans-1-isopropyl-3-methylcyclohexane

23. Glucose, shown below, is a simple sugar that consists of a six-membered ring. Draw glucose in its most stable conformation.

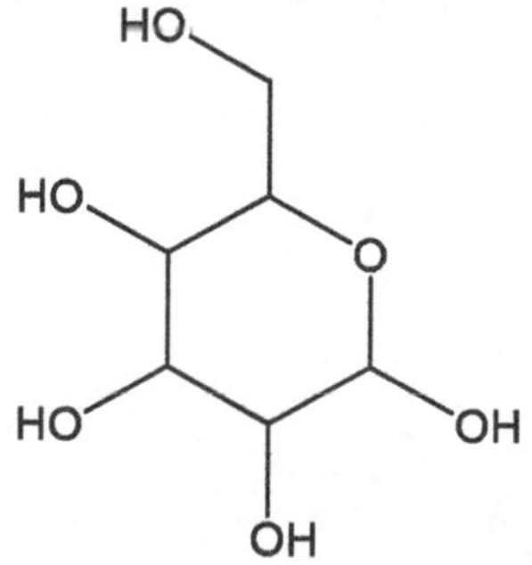

Class Activity 6A

Organic Reactions: Bond Breaking and Reactive Intermediates

Prior Knowledge:
Before beginning this activity, students should be familiar with the following concepts:

- Bonding.
- Definition of cations and anions.
- Potential energy diagrams.
- Curved arrow notation.

Learning Objectives
Content Learning Objectives:
After completing this activity students should be able to:
- Convey the difference between homolytic and heterolytic bond cleavage, using curved arrows to show the electron movement.
- Identify primary, secondary and tertiary cation, radicals and anions and determine their relative stability.
- Construct an energy diagram for a general organic reaction based on energy predictions for reactive intermediates.

Process Objectives:
- Information Processing. Students interpret information from an energy diagram and transfer the information to intermediate stability
- Critical Thinking. Students evaluate different types of intermediates to make predictions about organic reactions.

Class Activity 6A

Organic Reactions: Bond Breaking and Reactive Intermediates

Model 1: Heterolytic and Homolytic Bond Cleavage

When compounds react, bonds are broken and formed. Bond cleavage can be either heterolytic or homolytic.

Questions: For the reactions shown in Model 1:

1. (a)   Describe what happens to the electrons in the A-B bond during <u>heterolytic</u> cleavage.

   (b)   Explain what a full arrow (double barbed) illustrates.

2. (a)   Describe what happens to the electrons in the A-B bond during <u>homolytic</u> cleavage.

   (b)   Explain what a half arrow (single barbed or fishhook) depicts.

3. (a)   Ions are charged species. Which type of bond cleavage produces ions?
         (*circle one*) heterolytic / homolytic

   (b)   For the two intermediates produced in heterolytic bond cleavage, determine if they are electron rich or electron deficient.

                          A⊕ _____

                          B⊖ _____

   (c)   Recall that a positively charge species is a cation and a negatively charged species is an anion. Label the cations and anions of the intermediates below.

4. (a)   A radical species is a species that has an unpaired electron. Which type of bond cleavage produces radicals? (*Circle one*) heterolytic / homolytic.

    (b)   For the radical •CH₃, how many valence electrons are on the carbon? _____

    (c)   A radical species is (*circle one*) electron rich / electron deficient.

    (d)   Circle the radicals of the intermediates below.

$$\overset{\bullet}{\square} \qquad HO^{\ominus} \qquad H_3C\!:^{\ominus} \qquad H_3C^{\oplus} \qquad Br\bullet$$

5. For the reaction shown below:
    (a)   The type of bond cleavage that occurs is (*circle one*) heterolytic / homolytic.
    (b)   Identify the reactive intermediates as cation, anion or radical.
    (c)   Draw curved arrows to show electron movement for this bond cleavage.

$$\underset{CH_3}{\overset{CH_3}{H_3C-\underset{|}{\overset{|}{C}}-Br}} \longrightarrow \underset{CH_3}{\overset{CH_3}{H_3C-\underset{|}{\overset{|}{C}}\oplus}} + Br^{\ominus}$$

Model 2: Relative Stability of Reactive Intermediates

Intermediate	Primary (1°)	Secondary (2°)	Tertiary (3°)	Valence e-
Anion	$H_3C-\overset{..}{\underset{\ominus}{CH_2}}$	$H_3C-\underset{\ominus}{\overset{\overset{H}{\|}}{\overset{..}{C}}}-CH_3$	$H_3C-\underset{\ominus}{\overset{\overset{CH_3}{\|}}{\overset{..}{C}}}-CH_3$	
Cation	$H_3C-\underset{\oplus}{CH_2}$	$H_3C-\underset{\oplus}{\overset{\overset{H}{\|}}{C}}-CH_3$	$H_3C-\underset{\oplus}{\overset{\overset{CH_3}{\|}}{C}}-CH_3$	
Radical	$H_3C-\underset{\bullet}{CH_2}$	$H_3C-\underset{\bullet}{\overset{\overset{H}{\|}}{C}}-CH_3$	$H_3C-\underset{\bullet}{\overset{\overset{CH_3}{\|}}{C}}-CH_3$	

Questions:

6. (a)   For all of the intermediates in Model 2, circle the carbon containing the ⊕, ⊖, or •

    (b)   Consider the <u>primary</u> intermediates. How many carbons are directly attached to the carbon that is circled? _____

(c)   Consider the <u>secondary</u> intermediates. How many carbons are directly attached to the carbon that is circled? _____

(d)   Consider the <u>tertiary</u> intermediates. How many carbons are directly attached to the carbon that is circled? _____

(e)   Label each of the following intermediates as 1°, 2°, or 3°.

7. (a)   For each of the intermediates in Model 2, determine how many electrons are in the circled carbon's valence shell. (Fill in the last column of the table.)

(b)   Recall that carbon's preferred valence is 8 electrons. Which intermediates have <u>less</u> than 8 valence electrons? (*Circle all that apply*) cation / anion / radical.

Intermediates with less than 8 valence electrons would be expected to be (*circle one*) electron rich / electron deficient.

(c)   Which of the intermediates in Model 2 would be expected to be electron rich? (*Circle all that apply*) cation / anion / radical. Explain.

8. Alkyl groups donate electrons. From this information determine the following:
   (a)   Circle the <u>anion</u> that would be the most stable (lowest energy).   1° / 2° / 3°   Explain.

   (b)   Circle the <u>cation</u> that would be the most stable (lowest energy). 1° / 2° / 3°   Explain.

   (c)   Circle the <u>radical</u> that would be the most stable (lowest energy). 1° / 2° / 3°   Explain.

   (d)   The following list of cations is arranged from highest energy (least stable) to lowest energy (most stable). As a group, discuss whether this list is consistent with your conclusion in #8b. Explain.

   (least stable)   (most stable)
   ⊕H   ⊕CH₃   ⊕1°   ⊕2°   ⊕3°

Model 3: Energy Diagrams for a Reaction that Forms Intermediates

The energy diagram on the left represents the reaction A-B + C$\ominus$→A-C + B$\ominus$, while the energy diagram on the right represents just the <u>first step</u> of that reaction, A-B →A$\oplus$ + B$\ominus$.

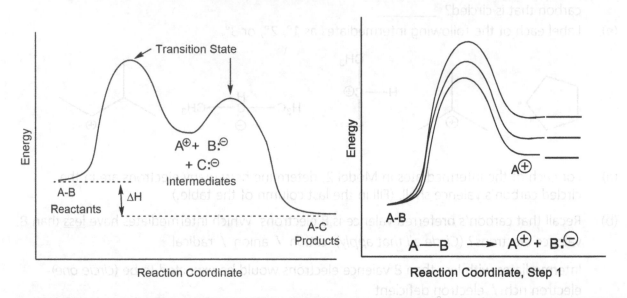

Questions:

9. For the energy diagram on the <u>left</u> side of Model 3:
    (a) The reactants are _____ and the products are _____. Circle the intermediates on the above diagram.
    (b) Which species are <u>highest</u> in energy? (*circle one*) reactants / intermediates / products. Explain how you came to this decision.

    (c) When proceeding from the reactants to the intermediates in the first step, energy is (*circle one*) consumed / released.

         The <u>first step</u> of the reaction is (*circle one*) endothermic / exothermic. Explain.

    (d) Once your group has reached agreement on the above questions, consider the overall reaction from the reactants to the products. The overall reaction is (*circle one*) endothermic / exothermic. Explain how you came to this decision.

10. For the energy diagram on the right side of Model 3:
   (a)   The reaction this energy diagram represents is shown below. Draw curved arrows to show electron movement for this reaction.

$$A\text{—}B \longrightarrow A^{\oplus} + B^{\ominus}$$

   (b)   What does A⊕ represent on the diagram?
   (*Circle one*) reactant / intermediate / product.

   (c)   If A⊕ represents three types of cations that can be formed after breaking the A-B bond, which cation would be the lowest in energy? (refer to #8d)
   (*Circle one*) 1° / 2° / 3°.

   (d)   On the lines to the right of the energy diagram above, label each energy level as 1°, 2° or 3°.

   (e)   Once your group has reached consensus on the above questions, rank the order of reactivity expected on formation of 1°, 2° or 3° cationic intermediates. (Rank from most reactive to least reactive). Explain.

11. (a)   Consider the formation of a <u>radical</u> intermediate. Based on your answer to #8c, what is the lowest energy radical intermediate that can be formed?
   (*Circle one*) 1° / 2° / 3°.

   (b)   Rank the order of reactivity expected on formation of 1°, 2° or 3° <u>radical</u> intermediates. (Rank from most reactive to least reactive). Explain.

   (c)   Consider the formation of an <u>anionic</u> intermediate. Based on your answer to #8a, what is the lowest energy anionic intermediate that can be formed?
   (*Circle one*) 1° / 2° / 3°.

   (d)   Rank the order of reactivity expected on formation of 1°, 2° or 3° anionic intermediates. (Rank from most reactive to least reactive). Explain.

12. Consider the following reactants.

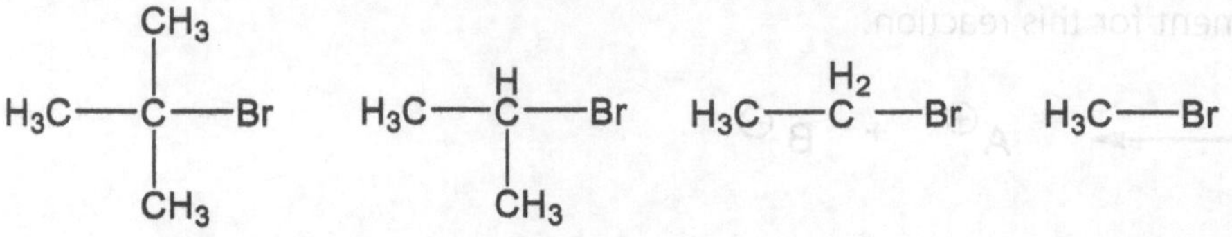

(a)   Which reactant would be most reactive when forming a cation intermediate? _____

(b)   Which reactant would be most reactive when forming a radical intermediate? _____

(c)   Which reactant would be most reactive when forming an anion intermediate? _____

Reflection: on a separate sheet of paper.
As a group, describe three concepts your group has learned from this activity and the one most important unanswered question about this activity that remains with your group. Turn this in before leaving class.

Additional Exercises:

13. Draw the products of <u>heterolytic</u> cleavage of the indicated bond for each of the compounds shown below. Draw curved arrows to show electron movement and label the products as anion, cation or radical.

(a)

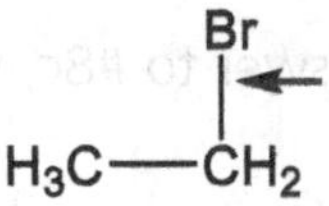

(b)

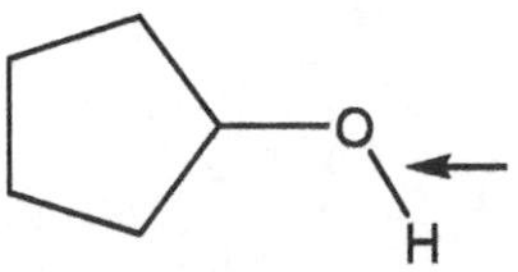

(c)

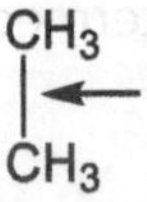

14. Draw the products of <u>homolytic</u> cleavage of the indicated bond for each of the compounds shown in #13 above. Draw curved arrows to show electron movement and label the products as anion, cation or radical.

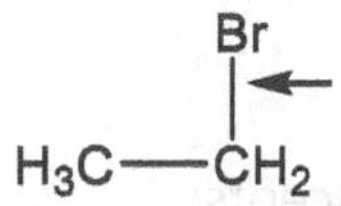

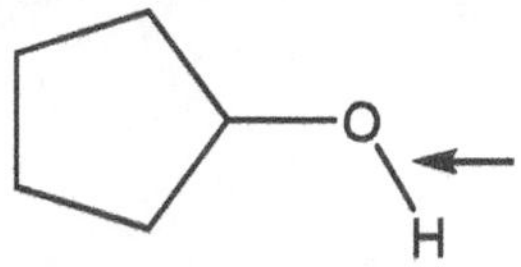

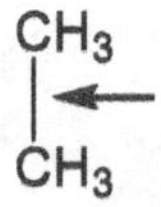

15. For each of the following pairs of intermediates, label as 1°, 2° or 3° and indicate which would be the lowest energy.
    (a)

    (b)

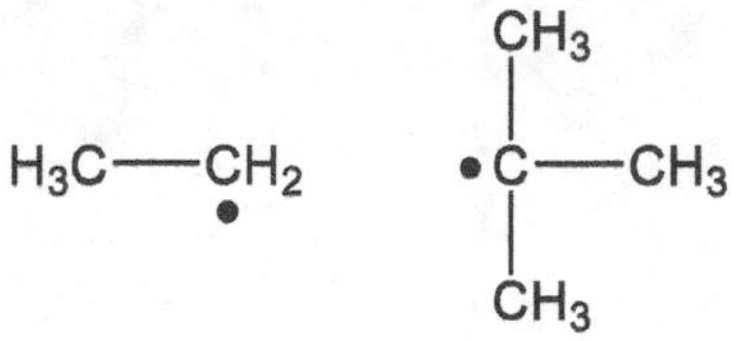

    (c)

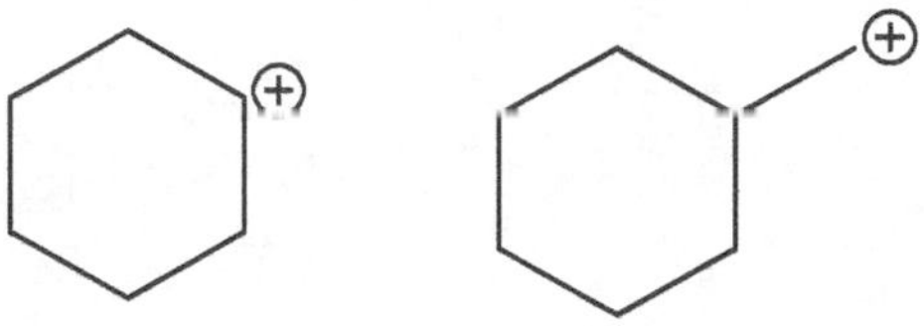

Class Activity 6B

Radical Halogenation Reactions

Prior Knowledge:
Before beginning this activity, students should be familiar with the following concepts:

- Homolytic bond cleavage.
- Relative stability of cations, anions and radicals.
- Potential energy diagrams.
- Curved arrow notation.

Learning Objectives
Content Learning Objectives:
After completing this activity students should be able to:
- Identify the steps in a radical chain mechanism.
- Draw the mechanism, using curved arrows, for a radical halogenation reaction.
- Predict which product would be most favored in a radical halogenation based on the stability of the radical intermediates.

Process Objectives:
- Critical Thinking. Students analyze and synthesize information about radical intermediates to reach a conclusion about radical bond formation.
- Problem Solving. Students execute a strategy to determine the steps of a radical chain mechanism.

Class Activity 6B

Radical Halogenation Reactions

Model 1: Radical Bond Formation

Homolytic bond cleavage gives radical intermediates as shown below.

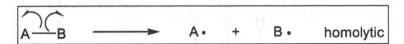

A radical can form a new bond by combining with another radical, or reacting with one electron of a sigma bond

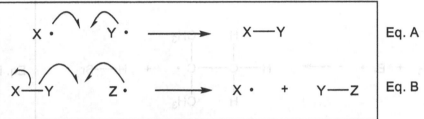

Questions:

1. For the reactions shown in Model 1:
   (a) Homolytic bond cleavage results in formation of (*circle one*) cations / anions / radicals.

      In homolytic cleavage, the electrons that make up the A-B single bond are divided up (*circle one*) equally / unequally between atoms A and B.

   (b) Once a radical is formed, it can quickly react to form other bonds. Describe how the bonds are forming in Eq. A.

   (c) Describe how the bonds are forming in Eq. B.

   (d) In which reaction are radicals formed as products? (*Circle one*) Eq. A / Eq. B.

   (e) In which reaction are radicals consumed so there are no radicals formed as products? (*Circle one*) Eq. A / Eq. B.

2. In addition to forming the products shown in Eq. B, the following alternate reaction could also take place. Show the mechanism, using curved half arrows, for this reaction. (HINT: compare to Eq. B in Model 1).

   $$Z\cdot \ + \ X\!-\!Y \ \longrightarrow \ Y\cdot \ + \ Z\!-\!X$$

Model 2: Reaction of Bromine Radical with Alkanes

Questions:

3. (a) In the reaction of a bromine radical with the alkanes in Model 2, what new bond is being formed? _____ In the product, which atom has the radical electron? _____

   (b) Label the carbon radicals from Model 2 as methyl, primary, secondary or tertiary.

   (c) Which reaction in Model 2 is the most likely? Explain.

4. (a) Add curved arrows to Eq. D and Eq. E to illustrate the mechanism of product formation.

(b)   Draw the products that would result if the following mechanism took place. Explain why this reaction is not likely to occur. (HINT: think of radical stability).

Model 3: Radical Chain Reactions

The mechanism (step by step pathway from reactants to products) of a radical halogenation reaction is a <u>chain reaction</u>. A chain reaction consists of the following three components:

1. **Initiation** – generation of radical
2. **Propagation** – radical reacts to form another radical
3. **Termination** – two radicals combine

Halogenation of an alkane follows the radical chain reaction mechanism.

Questions:

5. (a)   What is the <u>first</u> step of a radical chain reaction called?

(b)   Initiation involves generation of a radical. Are any radicals present in the reaction shown in Model 3? (*Circle one*) yes / no.

(c)   If no radicals are present, what has to happen chemically in order for the reaction to start? (See step 1 in Model 3.)

(d)   Radicals are formed via homolytic bond cleavage when Br_2 is exposed to heat or light. Use curved arrows (single barbed) to show this <u>initiation step</u> of the mechanism.

6. (a)   What is the <u>second</u> step of a radical chain reaction called?

(b)   What has to happen chemically in order to continue the radical reaction after initiation?

(c) Use curved arrows (single barbed) to show the <u>two</u> propagation steps leading to the product, CH_3Br. (HINT: not all Br_2 is consumed in the initiation step).

$$H{-}\overset{\overset{\displaystyle H}{|}}{\underset{\underset{\displaystyle H}{|}}{C}}{-}H \;+\; Br\cdot \longrightarrow H{-}\overset{\overset{\displaystyle H}{|}}{\underset{\underset{\displaystyle H}{|}}{C}}\cdot \;+\; H{-}Br$$

$$H{-}\overset{\overset{\displaystyle H}{|}}{\underset{\underset{\displaystyle H}{|}}{C}}\cdot \;+\; Br{-}Br \longrightarrow H{-}\overset{\overset{\displaystyle H}{|}}{\underset{\underset{\displaystyle H}{|}}{C}}{-}Br \;+\; Br\cdot$$

7. (a) What is the <u>third</u> step of a radical chain reaction called?

(b) What has to happen chemically in order to stop the radical reaction?

(c) Show two possible chain termination steps, where two radicals combine. (They do not have to lead to CH_3Br).

Model 4: Selective Halogenation

In the halogenation of propane, two products are formed in unequal amounts.

$$\wedge \;+\; Br_2 \xrightarrow{\text{light}} \underset{97\%}{\overset{Br}{\wedge\!\!\wedge}} \;+\; \underset{3\%}{\wedge\!\!\wedge Br}$$

$$\wedge \;+\; Cl_2 \xrightarrow{\text{light}} \underset{60\%}{\overset{Cl}{\wedge\!\!\wedge}} \;+\; \underset{40\%}{\wedge\!\!\wedge Cl}$$

Questions:

8. A selective reaction gives close to 100% of one product. An unselective reaction gives a mixture of products. Which reaction is selective? (*Circle one*) bromination / chlorination.

9. The major product obtained for either bromination or chlorination of propane is derived from a (*circle one*) primary / secondary / tertiary radical.

10. Explain why the major product in Model 4 is formed more readily than the minor product.

Reflection: on a separate sheet of paper.
As a group, describe three concepts your group has learned from this activity and the one most important unanswered question about this activity that remains with your group. Turn this in before leaving class.

Additional Questions:

11. Draw all the possible products that can be formed in the bromination of each of the following alkanes. Circle the product that will be formed in the greatest amount. (Only hydrogens attached to sp₃ hybridized carbons can be removed).

(a)

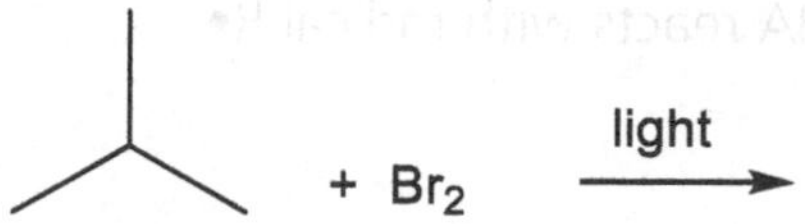

light

+ Br$_2$ ⟶

(b)

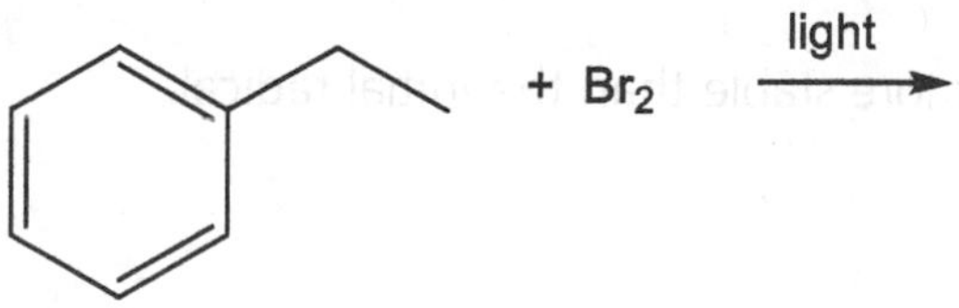

+ Br$_2$ light ⟶

(c).

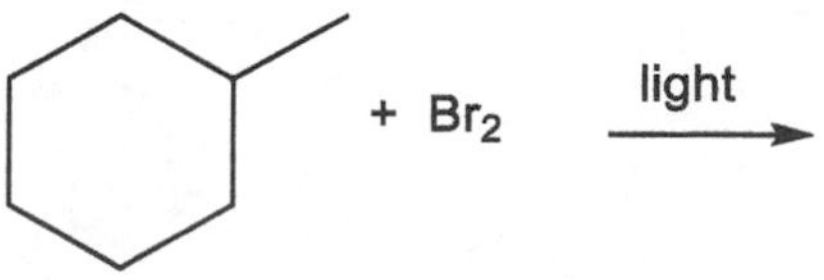

+ Br$_2$ light ⟶

12. Use curved arrows to draw the complete mechanism for the following radical mechanism.

13. Radical inhibitors react with any reactive radicals present in order to prevent further radical reactions from occurring. BHA (butylated hydroxyanisole) is added to foods to stop radical oxidations.

   (a)   Draw the radical that would be formed when BHA reacts with radical R•.

   (b)   Explain why this BHA radical might be more stable than the initial radical.

Class Activity 7

Stereochemistry

Prior Knowledge:
Before beginning this activity, students should be familiar with the following concepts:

- Isomers; constitutional, configurational and stereoisomers.
- Wedge/dash drawings.
- Atomic number.

Learning Objectives
Content Learning Objectives:
After completing this activity students should be able to:
- Identify a chirality center.
- Determine whether a compound is chiral or achiral and whether two compounds are enantiomers.
- Determine the absolute configuration of chiral compounds.

Process Objectives:
- Information Processing. Students manipulate 3-D models to visualize the 2-D representations of organic compounds.
- Critical Thinking. Students analyze chiral carbons to determine the absolute configuration based on defined processes.
- Teamwork. Students interact with each other to build molecular models and convince each other that the model matches the drawings on paper.

Class Activity 7

Stereochemistry

Model 1: Stereoisomers of Thalidomide

Thalidomide is important in medical history as it was prescribed to treat nausea and morning sickness in pregnant women. One stereoisomer of thalidomide caused birth defects including deafness, blindness and malformed limbs. <u>One</u> of the following compounds is thalidomide.

A $C_{13}H_{10}N_2O_4$ B $C_{13}H_{10}N_2O_4$

Definitions of key terms:
- **Stereoisomers** - have the same atom connectivity but cannot be interconverted by rotation around single bonds and differ only in the three dimensional arrangement of atoms in space.
- **Chirality center** (stereogenic center) – a carbon with four different groups.
- **Chiral compound** – has chiral center and no plane of symmetry; the mirror image is not the same as the original.
- **Achiral compound** – has plane of symmetry; the mirror image is the same as the original
- **Enantiomers** - non superimposable mirror images (cannot be aligned).

Questions: Consider compounds A and B in Model 1.
1. Do compounds A and B have the same molecular formula? (*Circle one*) yes / no.

2. A and B are (*circle one*) identical / constitutional / conformational isomers. Explain.

3. Put an asterisk next to any carbon(s) that have four different groups in both A and B. What term is used to identify this type of carbon (see list of terms)?

4. Determine if there is an internal plane of symmetry in compound A or B, and draw a dotted line to indicate where the plane is.

5. Based on your answer to #4, which of the above structures are <u>chiral</u> compounds?

6. Which of the above structures (if any) are <u>achiral</u> compounds?
7. Thalidomide is a chiral compound. Which structure, A or B is thalodimide?

8. Using wedge and dash structures at the chirality center, draw the two stereoisomers of thalodimide. Make sure everyone in the group agrees on these drawings.

9. The chirality center and four adjacent atoms have been extracted from Thalidomide to give the simplified structure below. Using a model kit, make two identical models of this molecule using blue for the NH_2 group and red for the C=O group. Confirm that they can be superimposed (aligned exactly) onto one another. Make sure that everyone in your group looks at the models and <u>check with an instructor</u> to make sure your model is correct.

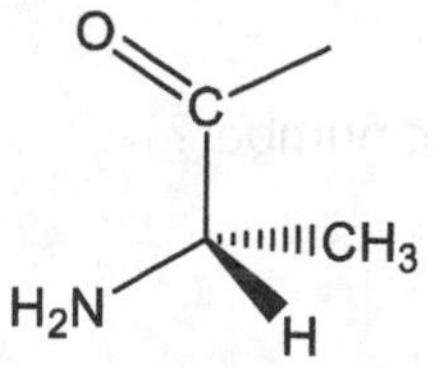

10. Switch any two groups on ONE of the models (leave the other model unchanged).
    (a)   Is this new model identical to the original model? (*circle one*) yes / no

    (b)   What definition from the following list best describes the relationship between the two models? (*Circle more than one answer if appropriate*) identical / stereoisomers / conformational isomers / constitutional isomers / enantiomers.

    (c)   Is the molecule chiral? Explain why.

Model 2: Naming Chirality Centers – R and S

Thalidomide exists in two stereoisomeric forms (enantiomers), but only one form causes birth defects. To distinguish between enantiomers, the R,S-system of naming (Cahn-Ingold-Prelog rules) was developed. This R,S assignment is the <u>absolute configuration</u> of that chiral center.

Questions:

11. Once the chiral center is identified, groups are prioritized, where 1 is the highest priority, and 4 is the lowest priority, as shown in Model 2.

    (a)   Which group in Model 2 has the highest priority? (*Circle one*) H / CH₃ / CH₂CH₃ / Br

    (b)   Which group in Model 2 has the lowest priority? (*Circle one*) H / CH₃ / CH₂CH₃ / Br

    (c)   Which atom from the groups in Model 2 has the highest atomic number?
           (*Circle one*) H / C / Br

    (d)   Which atom from the groups in Model 2 has the lowest atomic number?
           (*Circle one*) H / C / Br

    (e)   What conclusions can you make about priority ranking and atomic number?

12. To prioritize groups that have the same atom (in this case the two groups that have C), compare the next three groups connected to each carbon. This is illustrated using a tree diagram, shown below.

    (a)   The four atoms connected to the chiral C are drawn on the left. Label the highest priority group 1 and the lowest priority group 4.

(b)   Since the two C groups are a tie, the tree diagram shows what is bonded to each C. For the methyl group, the three atoms connected to the circled C are _____. For the ethyl group, the three atoms connected to the circled C are _____.

(c)   Compare these three atoms for each group. At the first point of difference, the group with the highest atomic number gets the higher priority. On the structure above, label the groups that get priority 2 and 3.

13. Once the priority of the groups has been established, the lowest ranking group must be oriented back or away from your perspective.

(a)   In Model 2, the molecule has been flipped to accomplish this. Next, trace a pathway from $1 \rightarrow 2 \rightarrow 3$. This is drawn in Model 2. This pathway is (circle one) clockwise / counterclockwise.

(b)   A chiral center is (R) if clockwise and (S) if counterclockwise. The chiral center in Model 2 is (circle one) R / S

14. For the following structure:

(a)   Rank each substituent on the chirality center of the molecule below from high priority (1) to low priority (4). If the priority cannot be determined based on the four atoms, draw tree diagrams to show the next set of adjacent atoms. (HINT: a double bond counts as being bonded to two of the same atoms).

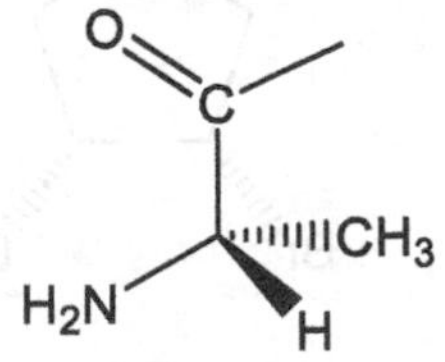

(b)   Rotate the compound and redraw it so that the low priority group is pointing back.

(c)   Indicate whether the pathway from high to low priority is clockwise or counterclockwise.

(d)   Indicate whether the compound is the (R) isomer or the (S) isomer.

15. Following the steps listed above, determine the absolute configuration, (R) or (S), of each chirality center in the compounds shown below.

(a)   (b)   (c)

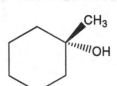 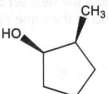 

Reflection: on a separate sheet of paper.
As a group, describe three concepts your group has learned from this activity and the one most important unanswered question about this activity that remains with your group. Turn this in before leaving class.

Additional Questions:

16. For each of the following compounds, identify any chirality centers and planes of symmetry. Determine whether the molecule is chiral.

(a)   (b)   (c)   (d)

17. For each of the following compounds:

(a)   (b)   (c)   (d)

(a)   Indicate with an asterisk any chirality centers in the molecules.
(b)   Rank each group on the chirality centers from high to low priority.
(c)   Determine whether the chirality center is (R) or (S). Remember that the lowest priority group MUST be oriented back.

18. Instead of redrawing the molecule to make the lowest priority group point back, alternatively, the lowest priority group can be swapped with the group that is currently pointing back. This is now the <u>enantiomer</u> of the original compound. This is illustrated in the questions below.

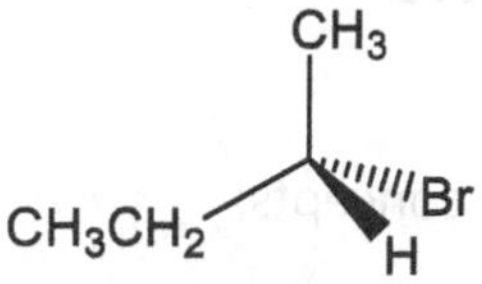

   (a)   Determine priorities of the groups attached to the chirality center.
   (b)   Swap the lowest priority group with the group that is pointing back, and redraw (this is the enantiomer).
   (c)   Determine the absolute configuration (R or S) of the enantiomer.
   (d)   Determine the absolute configuration (R or S) of the original compound.

19. Determine the absolute configuration for each of the following chiral drugs.
    (a) Ibuprofen (anti-inflammatory, anti pyretic)   (c) Albuterol (dilates airways)

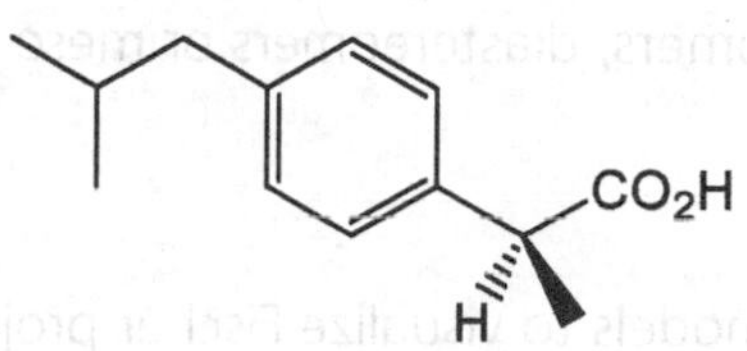

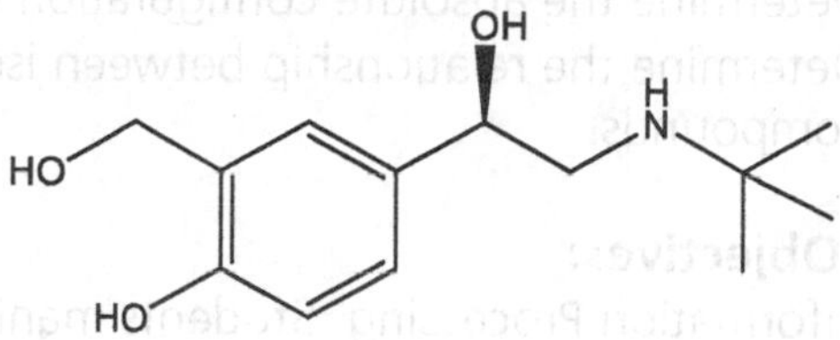

    (b) L-Dopa (used to treat Parkinson's disease)

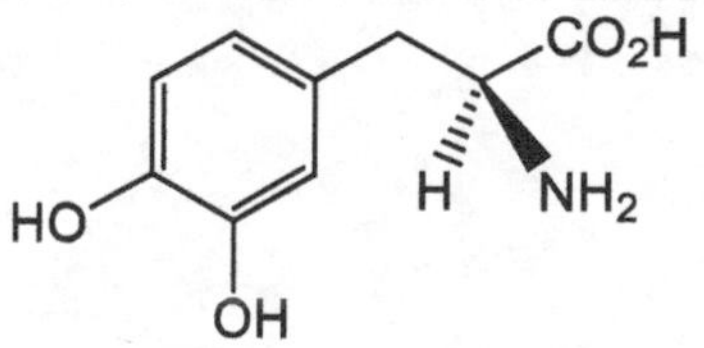

20. Two structures of Lipitor (a drug used to lower cholesterol) are shown below. Determine if they are the same compound or stereoisomers by determining the absolute configuration of each chiral center.

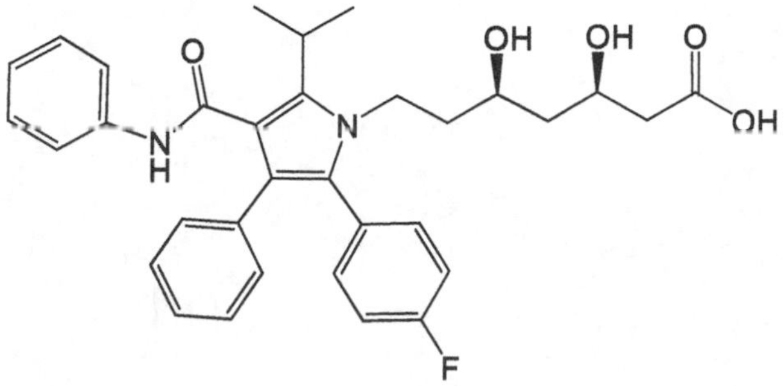

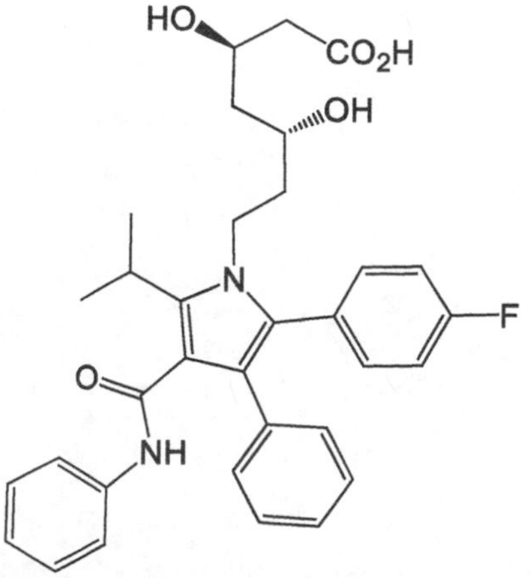

Class Activity 8

Stereoisomers and Fischer Projections

Prior Knowledge:

Before beginning this activity, students should be familiar with the following concepts:

- Isomers; constitutional, configurational and enantiomers.
- Wedge/dash drawings.
- R/S nomenclature.

Learning Objectives

Content Learning Objectives:

After completing this activity students should be able to:

- Interconvert wedge-dash, line angle and Fischer projections.
- Determine the absolute configuration of chiral carbons in a Fischer projection.
- Determine the relationship between isomers as enantiomers, diastereomers or meso compounds.

Process Objectives:

- Information Processing. Students manipulate the 3-D models to visualize Fischer projections and to interconvert different forms on paper.
- Teamwork. Students interact with each other to build models and convince each other whether the model matches the drawings on paper.

Class Activity 8

Stereoisomers and Fischer Projections

Model 1: Fischer Projections

CH₃ H►C◄Br CH₂CH₃ wedge dash	CH₃ H———Br CH₂CH₃ Fischer

Wedge dash $=$ Fischer (CH₃ top, H—C—Br, CH₂CH₃ bottom)

Questions:

1. Circle the carbon parent chain in the wedge dash structure in Model 1. In the Fischer projection the carbon chain is drawn (*circle one*) vertical / horizontal.

2. Compare the Fischer projection to the wedge dash structure in Model 1.
   (a) How is the Fischer projection different from the wedge dash model?

   (b) What direction is implied for vertical lines in the Fischer projection?

   (c) What direction is implied for horizontal lines in the Fischer projection?

3. Using your model kit, make a model of the wedge dash compound. As a group, compare the model to the Fischer projection shown. Discuss how the two structures represent the same compound.

4. Converting wedge-dash structures to Fischer Projections requires reorientation of the carbon framework so that the carbon chain is pointing <u>down</u>.
   (a) Make a model of the following compound in the staggered conformation. Rotate the C1-C2 bond 180° so that the carbon framework is in the eclipsed (U-shape) conformation. Redraw the rotated structure in wedge-dash form.

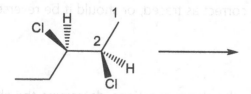

(b) Turn the model (keeping in the eclipsed conformation) so that the carbon chain is vertical, with C1 and C4 pointing down. Use wedge and dash lines only to draw this rotated form.

(c)   Convert the structure drawn to a Fischer Projection, using straight lines only.
(d)   Check with an instructor: the presenter should be prepared to explain how the group arrived at the above conclusions.

Model 2: Absolute Configuration in Fischer Projections

The absolute configuration (R or S) of a chiral center in a Fischer projection is determined in the same way as in a wedge dash structure. Recall that the lowest priority group (usually H) must be pointing back or away from you.

$$
\begin{array}{ccc}
& CH_3 & \\
H\!-\!\!\!&\!\!\!-\!Br & \equiv \qquad H\blacktriangleright C \blacktriangleleft Br \\
& CH_2CH_3 &
\end{array}
$$

Questions:
5. (a)   Rank the groups in the above Fischer projection from highest priority (1) to lowest priority (4).

(b)   In the Fischer projection what direction is the lowest priority group?

According to the rules, what direction does the lowest priority group need to be?

(c)   Trace a path from 1-2-3 based on your rankings above. The pathway is (*circle one*) clockwise / counterclockwise.

(d)   If the lowest priority group is in the opposite direction of where the rules state it should be, would the direction of the pathway be correct as traced, or should it be reversed? Explain.

(e)   Once your group has reached consensus on the above questions, determine the absolute configuration of the chirality center.

6. (a)   Swap the two horizontal groups in the Fischer projection above and redraw below.

   (b)   Determine the absolute configuration of the chirality center in the Fischer projection redrawn in #6a above.

   (c)   Compare your answers from #5e and #6b. What conclusions can you make about swapping two horizontal groups in Fischer projections?

7. (a)   Draw the (R) isomer of 2-bromo-1-butanol (shown below) using wedge dash structures.

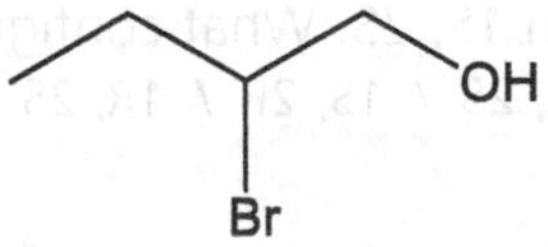

   (b)   Orient the structure drawn above so the carbon backbone is vertical with C1 at the top.

   (c)   Convert the structure in #7b to a Fischer projection.

   (d)   Check that your structure is correct by determining the absolute configuration of the chirality center of the Fischer projection.

Model 3: Stereoisomers

Stereoisomers - have same atom connectivity but cannot be interconverted by rotation around single bonds.
  (a)   Cis-trans isomers (geometric).
  (b)   Chiral (handedness) - a chiral object has a mirror image not identical to the original.
  - Enantiomers - non superimposable mirror images. If a compound has the R configuration, the enantiomer would have the S configuration (the exact opposite).
  - Diastereomers – compounds with two or more chirality centers are diastereomers if they are not mirror images. If a compound has the RR configuration (two chirality centers) then one diastereomer has the RS configuration.
  - Meso Compounds – compounds with two or more chirality centers that are achiral due to a plane of symmetry.

Questions:

8. Answer the following questions based on the definitions in Model 3.
  (a)   A compound has one chiral center with the R configuration. What configuration would the enantiomer have? (*Circle one*) R / S.

  (b)   A compound has one chiral center with the S configuration. What configuration would the enantiomer have? (*Circle one*) R / S.

  (c)   A compound has two chiral centers, with the configuration 1S, 2S. What configuration would the enantiomer have? (*Circle one*) 1S, 2S / 1S, 2R / 1R, 2S / 1R, 2R

  (d)   A compound has two chiral centers, with the configuration 1S, 2S. What configuration(s) would the diastereomer have? (*Circle as many as apply*) 1S, 2S / 1S, 2R / 1R, 2S / 1R, 2R.

9. (a)   Draw the mirror image of the following Fischer projection.

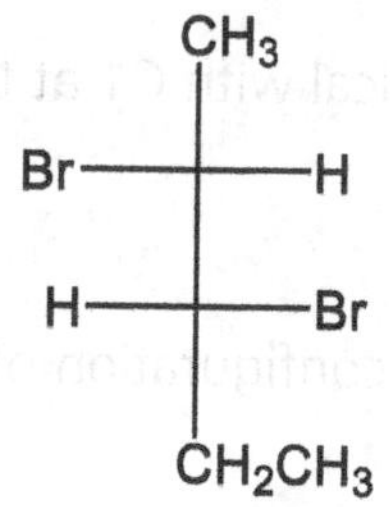

$$
\begin{array}{c}
CH_3 \\
Br \!\!-\!\!\!|\!\!-\!\! H \\
H \!\!-\!\!\!|\!\!-\!\! Br \\
CH_2CH_3
\end{array}
$$

  (b)   Determine the absolute configuration of each chiral center in the original _and_ mirror image.

  (c)   Based on your answer to #9b, determine if the mirror image is an enantiomer or the same as the original compound.

  (d)   Once your group has reached agreement on the above questions, describe how using the absolute configuration helps to determine the types of stereoisomers.

10. Refer to the definitions listed in Model 3 to determine whether the following pairs are enantiomers, diastereomers, or the same compound.

(a)   (b)

(c)   (d)

11. (a) Determine the absolute configuration for each chirality center of the structures shown in question #10a-d.

(b) Does this change your answers for question #10 on the relationship between each pair? Explain.

Reflection: on a separate sheet of paper.
As a group, describe three concepts your group has learned from this activity and the one most important unanswered question about this activity that remains with your group. Turn this in before leaving class.

Additional Questions:
12. Determine the absolute configuration of each chirality center, then indicate the relationship between the pairs of compounds as either the same, enantiomers, or diastereomers.

(a)   (b)

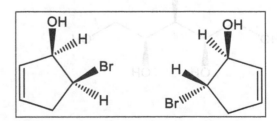

(c)

(d)

(e)

13. Fischer projections are typically used to represent carbohydrates, which contain multiple chirality centers. Determine the absolute configuration of each chiral center in the carbohydrate D-glucose shown below.

14. Xylitol is found in fruit and is used as a sugar substitute. Determine the absolute configuration of each chiral center, indicate if xylitol is a chiral compound and convert it to a Fischer projection.

Class Activity 9A

Substitution Nucleophilic Bimolecular, SN2

Prior Knowledge:
Before beginning this activity, students should be familiar with the following concepts:

- Bond polarity.
- Curved arrows.
- Energy diagrams.
- Lewis acid/base, nucleophile and electrophile definitions.

Learning Objectives
Content Learning Objectives:
After completing this activity students should be able to:
- Identify a substitution reaction and label the species acting as nucleophile and electrophile.
- Draw the mechanism for an SN2 reaction and label on an energy diagram.
- Determine the relative strength of nucleophiles and electrophiles and use this information to predict the direction of an SN2 reaction.

Process Objectives:
- Information Processing. Students interpret information about nucleophiles and electrophiles and the roles in a substitution reaction.
- Critical Thinking. Students analyze the substitution reaction mechanism to predict the direction of the reaction and the final product.

Class Activity 9A

Substitution Nucleophilic Bimolecular, SN2

Model 1: Nucleophilic Substitution Reactions
In a substitution reaction, a nucleophile reacts with a compound containing a leaving group.

Questions:

1. For the general substitution reaction (Eq. 1) in Model 1:
   (a)   What bond is formed in this reaction?
   (b)   What bond is broken in this reaction?
   (c)   Circle the leaving group, or the atom that leaves from the reactant.
   (d)   Provide an explanation for why this reaction is called a <u>substitution</u> reaction.

2. (a)   A nucleophile (nucleus loving) is attracted to an atom with (*circle one*) ⊕ or ⊖ charge.
   (b)   A nucleophile will (*circle one*) donate / accept electrons. Based on this answer, would a nucleophile be considered a Lewis Acid or a Lewis Base?
   (c)   Consider the C-X bonds of the reagents in Model 1, where X is a halogen. Add a δ+ and δ- to these bonds to indicate the direction the C-X bond is polarized.
   (d)   What part of the C-X bond will the nucleophile be attracted to? (*Circle one*) C / X

   (e)   An electrophile (electron loving) is attracted to an atom with (*circle one*) ⊕ / ⊖ charge.
   (f)   An electrophile has electrons to (*circle one*) donate / accept. Based on this answer, would an electrophile be considered a Lewis Acid or a Lewis Base?
   (g)   Label the nucleophile and electrophile in Eq. 2 and Eq. 3 in Model 1.

3. Use curved arrows above to illustrate the mechanism that will accomplish each substitution reaction shown in Model 1. (Remember that curved arrows show electron movement.)

Model 2: One-Step Nucleophilic Substitution (SN2)

The mechanism for an SN2 reaction is shown below. The highest potential energy point between the starting materials and the product is the transition state.

transition state

Questions:

4. For the reaction shown in Model 2:
   (a) How many reactants are involved in the reaction? _____
   (b) Label the nucleophile (Nu⊖) and electrophile (E⊕), and circle the leaving group.
   (c) What bond is formed in this reaction? _____ What bond is broken? _____
   (d) Draw curved arrows to show electron movement.

5. Focus on the transition state for the reaction shown in Model 2:
   (a) How many of the reactants are present in the transition state?

   (b) Based on your answer for #5a, explain why an SN2 reaction is considered **bimolecular**.

   (c) The rate equation for the SN2 reaction is Rate = [CH_3Br][I⊖]. Explain how the rate equation supports a **bimolecular** transition state.

   (d) The dotted line in the transition state represents a partial bond. This suggests that the bonds are being formed and broken (*circle one*) stepwise / simultaneous? Explain.

   (e) Based on your answer for #5d, explain why the SN2 reaction is a **concerted** reaction. (concerted means performed in unison).

   (f) In the transition state the incoming group (Nu⊖) is on the (*circle one*) same / opposite side of the leaving group.

(g) Determine why the transition state shown in Model 2 would be more favorable than a transition state where the incoming group and the leaving group are on the same side.

6. Draw a potential energy diagram to illustrate the reaction mechanism shown in Model 2. Label with reactants, transition state and product structures.

7. Focus on the stereochemistry of the following reaction.

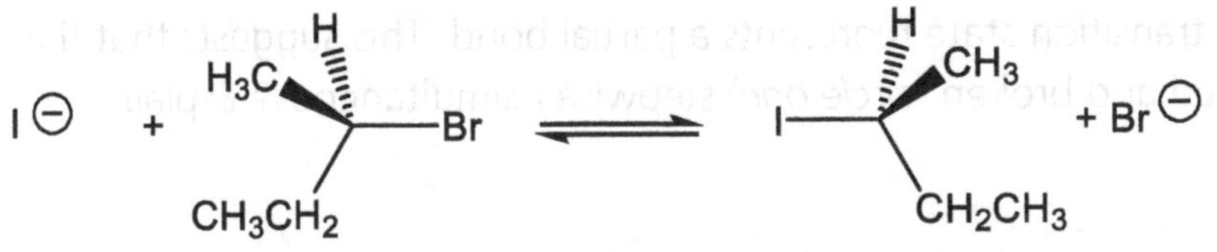

$$I^{\ominus} \; + \; \underset{CH_3CH_2}{\overset{H_3C\quad H}{C}}-Br \; \rightleftharpoons \; I-\underset{CH_2CH_3}{\overset{H\quad CH_3}{C}} \; + \; Br^{\ominus}$$

(a) Determine the absolute configuration for the reactant and the product.

(b) Explain how the stereochemistry proves which side the nucleophile approaches the reactant with reference to the leaving group.

Model 3: Nucleophiles

Recall that a nucleophile (Nu⊖) is a Lewis base which donates electrons.

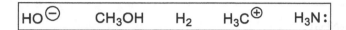

$$HO^{\ominus} \qquad CH_3OH \qquad H_2 \qquad H_3C^{\oplus} \qquad H_3N:$$

Questions:

8. For the compounds listed in Model 3:
   (a)   Circle the compounds that could act as nucleophiles, cross out those that would not.
   (b)   Explain why the compounds circled could act as nucleophiles.

   (c)   What feature must all nucleophiles have?

   (d)   Once your group has reached consensus, come up with an explanation of why the crossed out compounds would not act as nucleophiles.

9. Circle the compound in each of the following pairs that is the better nucleophile. Explain your choice using terms such as charge, electronegativity, size, basicity and polarizability.
   (a)   CH₃OH   vs   CH₃O⊖

   (b)   CH₃⊖   vs   NH₂ ⊖

   (c)   HO⊖   vs   F⊖

   (d)   I⊖   vs   Cl⊖

   (e)

   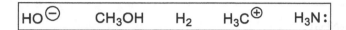

   vs

Model 4: Electrophiles
Recall that an <u>electrophile</u> (E⊕) is a Lewis Acid which accepts electrons.

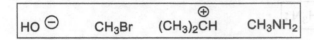

Questions:
10. For the compounds listed in Model 4:
    (a)   Circle the compounds that could act as electrophiles, cross out those that would not.
    (b)   Explain why the compounds circled could act as electrophiles.

    (c)   What feature must all electrophiles have?

    (d)   Once your group has reached consensus, come up with an explanation of why the crossed out compounds would not act as electrophiles.

11. Consider the following reaction:

$$R\!-\!X \rightleftharpoons R^{⊕} + X^{⊖}$$

    (a)   Label the RX bond with δ+ and δ- to indicate bond polarity.
    (b)   Why would RX be considered to be an electrophile?

12. Consider the acid dissociation of HX below:

$$H\text{-}X \rightleftharpoons H^{⊕} + X^{⊖}$$

    (a)   If X⊖ is stable, is HX likely to accept an electron pair, to form products? Explain.

    (b)   If X⊖ is a strong base, is HX likely to accept an electron pair, to form products? Explain.

    (c)   If X⊖ is a weak base, is HX likely to accept and electron pair, to form products? Explain.

(d)   The pKa of HI = -10, and for CH_3OH = 16. Use these values to determine which direction the following reaction will proceed. Explain your answer based on the strength/stability of the Lewis Base.

$$CH_3CH_2\text{-}I \quad + \quad {}^{\ominus}OCH_3 \quad \overset{?}{\rightleftharpoons} \quad CH_3CH_2OCH_3 \quad + \quad I^{\ominus}$$

13. Circle the better leaving group in each of the following pairs. Explain your answer using terms such as charge, electronegativity, size, basicity and polarizability.

   (a) $I^{\ominus}$ vs $Cl^{\ominus}$

   (b) $Br^{\ominus}$ vs $F^{\ominus}$

   (c) $HO^{\ominus}$ vs $Cl^{\ominus}$

   (d) $HO^{\ominus}$ vs H_2O

Reflection: on a separate sheet of paper.

As a group, describe three concepts your group has learned from this activity and the one most important unanswered question about this activity that remains with your group. Turn this in before leaving class.

Additional Questions:

17. Steric hindrance occurs when bulky groups become too close to one another, causing unfavorable electron repulsion.

   (a)   Draw the structures of bromoethane and 2-bromopropane. Label as 1°, 2° or 3°.

(b) Given that steric hindrance is unfavorable, predict whether the SN2 reaction of bromoethane or 2-bromopropane would be faster. Explain. (HINT: draw the transition state for each reaction).

(c) Based on the answers to the above questions, determine the order of reactivity of alkyl halides in an SN2 reaction (methyl, 1° etc.)

17. Predict the product of each of the following reactions, showing stereochemistry if appropriate. Identify the nucleophile and electrophile.

(a)

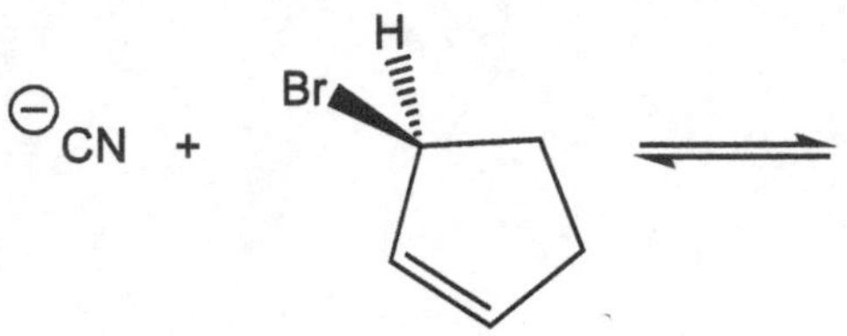

(b)

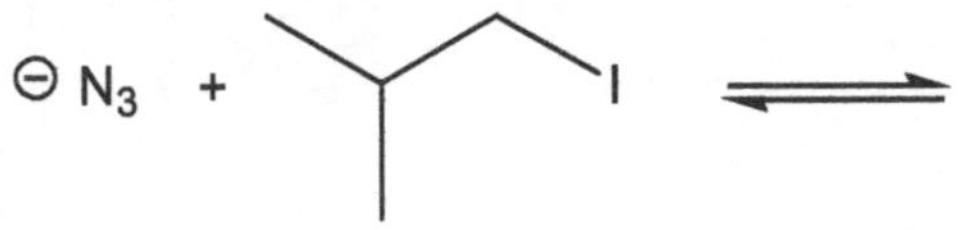

18. SN2 reactions are useful in the synthesis of many pharmaceutical compounds. The analgesic reagent ibuprofen can be prepared using an SN2 reaction as outlined below in the box. Draw the reagent that would be needed to accomplish this SN2 reaction, and provide a mechanism for the reaction.

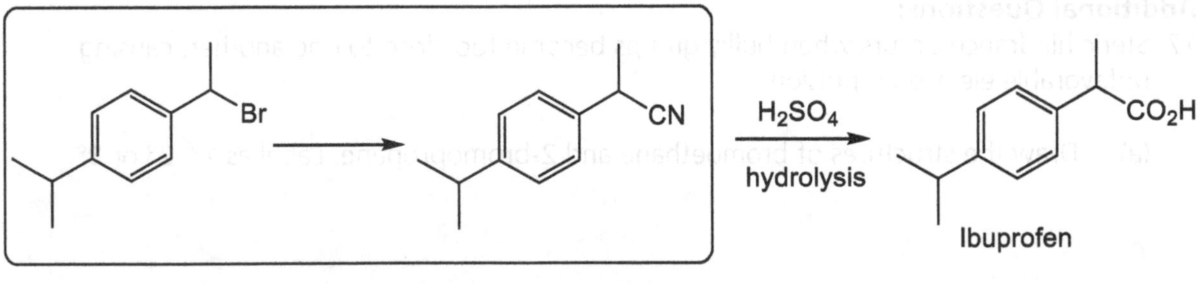

Class Activity 9B

Substitution Nucleophilic Unimolecular, SN1

Prior Knowledge:
Before beginning this activity, students should be familiar with the following concepts:

- Bond polarity.
- Curved arrows.
- Energy diagrams.
- Lewis acid/base, nucleophile and electrophile definitions.
- Substitution reaction, SN2 mechanism.

Learning Objectives
Content Learning Objectives:
After completing this activity students should be able to:
- Draw the mechanism for an SN1 reaction and label on an energy diagram.
- Rank the stability of carbocation intermediates and use to determine the reactivity in SN1 vs SN2 mechanisms.
- Determine which compounds will undergo rearrangements and draw the resulting products that occur on rearrangement.

Process Objectives:
- Information Processing. Students interpret information about cation stability and starting material reactivity.
- Critical Thinking. Students analyze and compare the SN1 and SN2 mechanisms, and predict which mechanism will occur with a given set of reagents.

Class Activity 9B

Substitution Nucleophilic Unimolecular, SN1

Model 1: Two-Step Nucleophilic Substitution (SN1)

SN1 (two-step) mechanism

Questions:

1. For the reaction shown in Model 1:
   (a) Step 1 occurs (*circle one*) slow / fast. Step 2 occurs (*circle one*) slow / fast.
   (b) What bond is broken in <u>Step 1</u> of the SN1 reaction?

   (c) What bond is formed in <u>Step 2</u> of the SN1 reaction?

   (d) Label the nucleophile (Nu⊖) and the electrophile (E⊕) for each step, and circle the leaving group. Draw curved arrows to show electron movement.

   (e) Deprotonation (final step) involves loss of the proton that was part of the incoming group, to return the oxygen to a neutral charge. Draw curved arrows to illustrate this proton loss.

2. The mechanism for an SN1 reaction occurs in two steps as shown in Model 1:
   (a) As a group, discuss why the first step is slow and the second step is fast.

   (b) Which step would be the rate determining step? (*Circle one*) Step 1 / Step 2.

   (c) How many reactants are involved in the rate determining step?

   (d) Based on your answer to #2c above, explain why the SN1 reaction is **unimolecular**.

(e) The rate equation for an SN1 reaction is Rate = k[CH_3Br]. In your groups discuss how the rate equation supports a <u>unimolecular</u> reaction. Explain your answer.

Model 2: Carbocation Intermediates

Shown below is the first step of an SN1 reaction, formation of a carbocation intermediate.

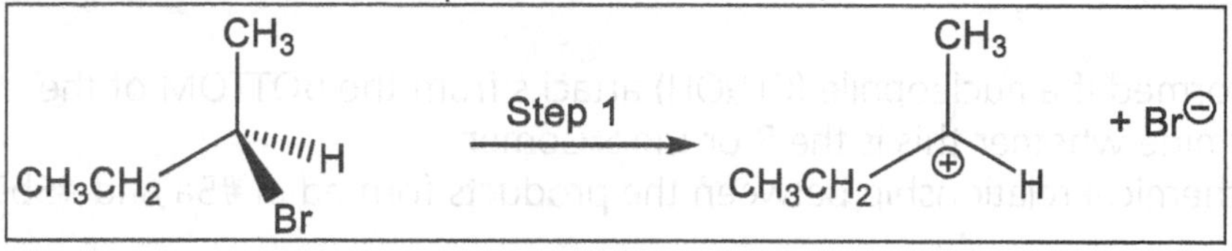

Questions:

3. (a) What is the hybridization of the carbon with the ⊕ charge in the carbocation intermediate? (*Circle one*) sp / sp$_2$ / sp$_3$

    (b) What is the shape of the carbocation? (*Circle one*) tetrahedral / trigonal planar

    (c) How many valence electrons does the carbon with the ⊕ charge have? _____

    (d) The carbocation is (*circle one*) electron rich / electron deficient.

    (e) The carbocation is most likely to react with electron (*circle one*) rich / deficient reagents.

4. (a) Rank the following carbocations from most stable (1) to least stable (4). (Recall from CA6A that alkyl groups donate electrons).

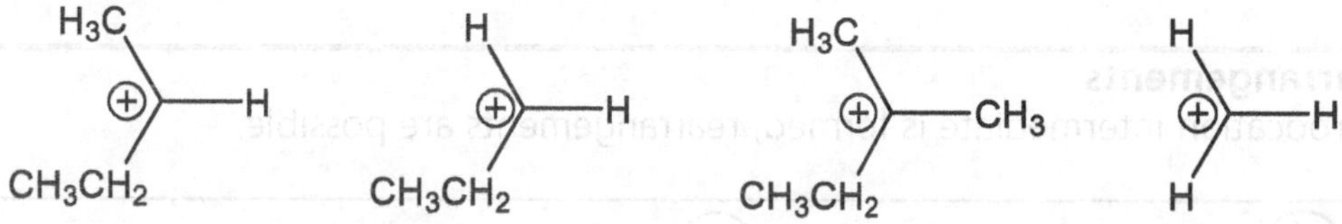

    (b) Which cation type will have the lowest energy? (*Circle one*) 1° / 2° / 3°.

    (c) Based on the answers to the above questions, determine the order of reactivity of alkyl halides in an SN1 reaction (methyl, 1° etc.)

    (d) In the space below, draw a potential energy diagram that illustrates the <u>entire</u> two-step SN1 reaction mechanism that shows conversion of RX → R+ → RNu.

5. Starting with the carbocation shown in Model 2:
   (a) Draw the product formed if a nucleophile (CH_3OH) attacks from the TOP of the carbocation. Determine whether this is the R or the S isomer.

$$CH_3CH_2 - \overset{\displaystyle CH_3}{\underset{\displaystyle Br}{C}} \text{'''''''} H \longrightarrow$$

   (b) Draw the product formed if a nucleophile (CH_3OH) attacks from the BOTTOM of the carbocation. Determine whether this is the R or the S isomer.
   (c) What is the stereochemical relationship between the products formed in #5a and #5b?

   (d) Retention of configuration means the incoming group replaced the leaving group on the <u>same</u> side. Which product shows retention of configuration? (*Circle one*) #5a / #5b
   (e) Inversion of configuration means the incoming group replaced the leaving group on the <u>opposite</u> side. Which product shows inversion of configuration? (*Circle one*) #5a / #5b
   (f) Once your group has reached agreement on the above questions, determine if the products from #5a and #5b above would be formed in equal amounts or not. If not in equal amounts, which product would be favored? Explain your answer.

Model 3: Rearrangements
Any time a carbocation intermediate is formed, rearrangements are possible.

$$H - \overset{\oplus}{C} - \overset{\overset{\displaystyle \textcircled{H}}{|}}{C} - CH_3 \quad \xrightarrow[\substack{\text{shift} \\ \text{FAST}}]{\text{1,2-hydride}} \quad H - \overset{\overset{\displaystyle \textcircled{H}}{|}}{C} - \overset{\oplus}{C} - CH_3 \quad \xrightarrow{Nu^{\ominus}} \quad H - \overset{\overset{\displaystyle H}{|}}{C} - \overset{\overset{\displaystyle Nu}{|}}{C} - CH_3$$

Cation A   Cation B

Questions:
6. For the rearrangement reaction in Model 3:
   (a) Label the cations A and B as 1°, 2° or 3°. Which cation is more stable? (*Circle one*) A / B.

   (b) In the rearrangement of cation A to cation B, what bond is being broken?

What bond is being formed?

(c)   Explain what the curved arrow represents in showing the rearrangement.

(d)   Hydride ion is represented as [H:⊖]. In your groups discuss why this rearrangement is called a <u>hydride</u> shift and NOT a <u>proton</u> shift?

(e)   Once your group has reached consensus on the above questions, determine why the rearrangement from cation A to B is fast. Would the reverse (B to A) be favored as well? Explain.

7. (a)   Draw the carbocations formed from migration of Ha, and from migration of Hb.

(b)   Identify all the carbocations (the initial cation as well as the cations formed on rearrangement of Ha and Hb) as 1°,2°,or 3°.

(c)   Which hydrogen will be more likely to migrate? (*Circle one*) Ha / Hb. Explain.

Model 4: SN1 vs SN2 Mechanism

$$\text{(structure)} \quad Br + CH_3O^\ominus \xrightarrow{Et_2O} \text{(structure)} OCH_3 + Br^\ominus \quad SN2$$

$$\text{(structure)} Br + CH_3OH \xrightarrow{CH_3OH} \text{(structure)} OCH_3 + Br^\ominus \quad SN1$$

Questions:

8. For the reaction labeled <u>SN2</u> in Model 4:
    (a) The alkyl halide is (*circle one*) 1° / 2° / 3°. What type of alkyl halide reacts fastest in an SN2 reaction? Explain.

    (b) The nucleophile is (*circle one*) strong / weak.

    (c) For SN2 reactions, Rate=k[RX][Nu]. Is the concentration or strength of the nucleophile important in an SN2 reaction? Explain.

    (d) The solvent is polar (*circle one*) protic / aprotic. Would this solvent be capable of hydrogen bonding with the nucleophile? _____ How would this affect the strength of the nucleophile?

9. For the reaction labeled <u>SN1</u> in Model 4:
    (a) The alkyl halide is (*circle one*) 1° / 2° / 3°. What type of alkyl halide reacts fastest in an SN1 reaction? Explain why.

    (b) The nucleophile is (*circle one*) strong / weak.

    (c) For SN1 reactions, Rate=k[RX]. Is the concentration or strength of the nucleophile important in an SN1 reaction? Explain.

    (d) The solvent is polar (*circle one*) protic / aprotic. Would this solvent be capable of hydrogen bonding with the leaving group? _____ Would the solvent be capable of stabilizing the cationic intermediate through dipole interactions? _____ How would these factors affect the stability of the intermediate?

POGIL
WWW.POGIL.ORG
Copyright © 2015

10. Once your group has reached consensus on the above questions, predict whether an SN1 or SN2 mechanism will occur with the reaction conditions indicated. Complete the table, providing an example of the reaction condition when applicable, and indicate which mechanism is most probable under this condition. Explain.

Reaction Condition	Example	Mechanism (SN1 or SN2)
High concentration of Nu: $\ominus$		
Low concentration of Nu: $\ominus$		
Weak Nu: $\ominus$		
Strong Nu: $\ominus$		
Polar aprotic solvent		
Polar protic solvent		

Reflection: on a separate sheet of paper.

As a group, describe three concepts your group has learned from this activity and the one most important unanswered question about this activity that remains with your group. Turn this in before leaving class.

Additional Questions:

11. Determine the mechanism for each of the following reactions as either SN1 or SN2, and draw the resulting product.

    (a)

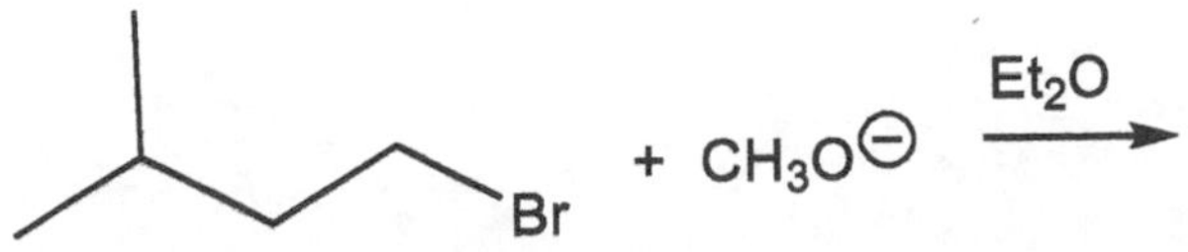

$+ \ CH_3O^{\ominus} \xrightarrow{\text{Et}_2O}$

POGIL
WWW.POGIL.ORG
Copyright © 2015

(b)

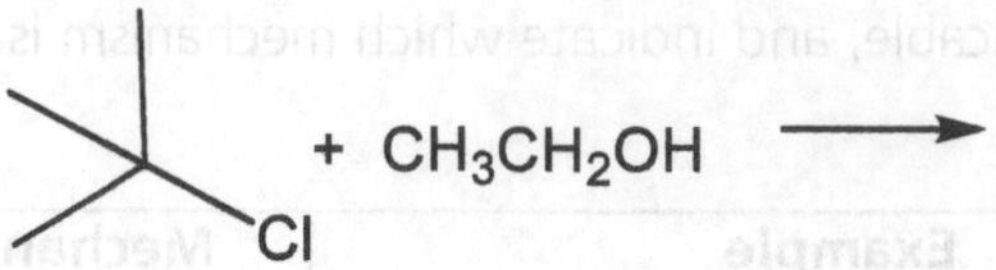

+ CH_3CH_2OH ⟶

(c)

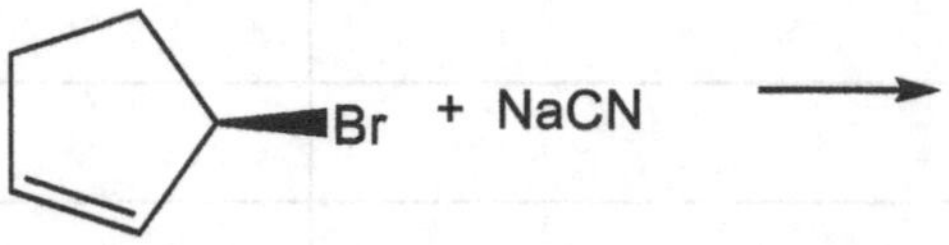

+ NaCN ⟶

12. Rationalize the formation of the following product (show mechanism).

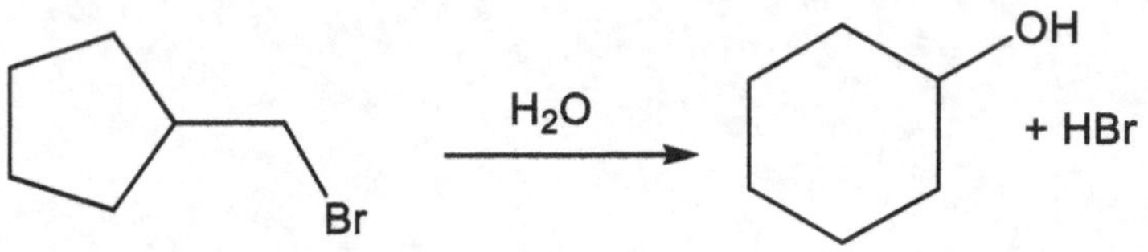

$\xrightarrow{H_2O}$

+ HBr

13. Reactions that occur via the SN1 mechanism are not as useful in the synthesis of pharmaceutical reagents. Rationalize why this is so.

Class Activity 10A

Elimination

<u>**Prior Knowledge:**</u>
Before beginning this activity, students should be familiar with the following concepts:

- Functional groups and naming organic compounds.
- Acid / base roles and definitions.
- Curved arrow notation.
- Rate equations.
- Substitution reactions, SN1 and SN2 mechanism (only for Q#14 and Q#15).

<u>**Learning Objectives**</u>
Content Learning Objectives:
After completing this activity students should be able to:
- Identify α, β and γ positions in an alkyl halide.
- Predict and draw the major elimination product based on Zaitzev's rule.
- Draw the mechanism using curved arrows for both E1 and E2 mechanisms.

Process Objectives:
- Information Processing. Students interpret information about the elimination reaction, mechanism and selectivity.
- Critical Thinking. Students evaluate the possible elimination mechanisms and predict the major products of the reaction.

Class Activity 10A

Elimination

Model 1: Elimination

Questions:
1. For the reaction shown in Model 1:
   (a) What is the functional group of the reactant?

   What is the functional group of the organic product?

   (b) What two atoms in the reactant are eliminated to form the organic product?

2. (a) Provide the IUPAC name for the reactant.

   (b) Look at the atoms that were eliminated. On what carbons (indicate the carbon numbers) were those atoms located?

   (c) Greek characters can also be used to locate groups instead of numbers. Carbons 1, 2, 3 would be α, β, γ. On the reactant in Model 1, label the carbon containing the Br as α. Continue labeling β and γ carbons.

   (d) For the groups that are eliminated, indicate where they are located using Greek characters.

   (e) Elimination is often called β–elimination. Explain what this means.

3. Label the α carbon and β–carbons in each of the following compounds. Draw the β–elimination product expected from each compound.

4. In each of the above examples, there is only one β–elimination product expected. Once your group has reached agreement on the above questions, determine why there would not be two products expected for the first two compounds shown in #3.

Model 2: Elimination with Multiple β–Carbons

Questions:

5. For the alkyl halide shown in Model 2:
    (a)   Label the α carbon and the <u>two</u> β carbons.

    (b)   Put an A next to the hydrogen that when eliminated will form alkene A, and a B next to the hydrogen that when eliminated will form alkene B.

6. Alkenes that are more substituted are more stable.
    (a)   Draw in the two carbons that make up the double bond in compound A. In total how many R groups (NOT hydrogen) are attached to these <u>two</u> carbons?

    (b)   Draw in the two carbons that make up the double bond in compound B. In total how many R groups are attached to these two carbons?

    (c)   Which alkene would be the most stable? *(Circle one)* A / B.

7. (a)   Once your group has reached agreement on the above questions, determine which compound, A or B would be the major product. Explain your answer.

    (b)   Zaitsev's Rule states that the more substituted alkene will be the major elimination product. Does your answer to #7a agree with Zaitsev's Rule? *(Circle one)* yes / no. Explain.

8. For each of the following compounds, label the α and β carbons, draw all possible elimination products and circle the major product that will form.

Model 3: Mechanism of Elimination, Bimolecular (E2)

The mechanism for an E2 reaction is shown below. Recall that the highest potential energy point between the starting materials and the product is the transition state.

transition state

Questions:

9. For the E2 mechanism shown in Model 3:
    (a) The base, B⊖, removes a hydrogen to form BH. What position is the hydrogen relative to the Br group? (*Circle one*) α / β / γ

    (b) What bonds are being formed in this reaction?

    What bonds are being broken in this reaction?

    (c) How many steps are involved in the E2 mechanism?

10. Focus on the transition state for the reaction shown in Model 3:
    (a) How many reactants are involved in the reaction? _____ How many of the reactants are present in the transition state? _____

(b)   Based on your answer for #10a, explain why the E2 reaction is considered **bimolecular**.

(c)   The rate equation for the E2 reaction is Rate = [RBr][BΘ]. Explain how the rate equation supports a **bimolecular** transition state.

(d)   Is the strength of the <u>base</u> important in an E2 mechanism? (*Circle one*) yes / no. (HINT: is the rate dependent on the base?)

(e)   The dotted line in the transition state represents a partial bond. This suggests that the bonds are being formed and broken (*circle one*) stepwise / simultaneous (concerted). Explain.

(f)   Once your group has reached agreement on the above questions, compare the E2 mechanism to the SN2 mechanism. Describe and similarities and/or differences.

11. Given the following key points:
   • Steric effects are minimal in an E2 mechanism
   • A more substituted alkene is more stable than a less substituted alkene
      Determine the order of reactivity of alkyl halides (1°, 2 °, 3 °) in the E2 mechanism. Explain.

Model 4: Mechanism of Elimination, Unimolecular (E1)

Questions:

12. For the E1 reaction mechanism shown in Model 4:
    (a)   How many steps are in the E1 mechanism? _____
    (b)   What bonds are formed and/or broken in <u>Step 1</u> of the E1 reaction?

    (c)   What bonds are formed and/or broken in <u>Step 2</u> of the E1 reaction?

    (d)   Draw curved arrows to show electron movement for both steps of the reaction.

13. For the E1 mechanism:
    (a)   Which step would be the rate determining step? (*Circle one*) Step 1 / Step 2.

    (b)   How many reactants are involved in the rate determining step?

    (c)   Based on your answer to #13b above, explain why the E1 reaction is unimolecular.

    (d)   The rate equation for an E1 reaction is Rate = k[RBr]. In your groups discuss how the rate equation supports a unimolecular reaction.

    (e)   Is the strength of the <u>base</u> important in an E1 mechanism? (*Circle one*) yes / no. (HINT: is the rate dependent on the base?)

    (f)   In the presence of a <u>strong</u> base, discuss in your groups whether an alkyl halide would react via the E1 or the E2 mechanism. Explain your rationalization.

14. A carbocation intermediate is formed in the first step of an E1 mechanism.
    (a)   Based on the stability of the intermediate, which class of alkyl halide would be most reactive via the E1 mechanism? (*Circle one*) Methyl / 1° / 2° / 3 °.

    (b)   Compare the first step of the SN1 mechanism (CA9B) with the first step of the E1 mechanism. Note any similarities and/or differences.

    (c)   Indicate whether CH_3OH is a (*circle one*) strong / weak base.
            Indicate whether CH_3OH is a (*circle one*) strong / weak nucleophile.

    (d)   Given your answer to #14c above, determine whether the reaction of t-butyl bromide in CH_3OH would give SN1, E1 or mixtures of both. Explain.

15. (a) Recall that an SN1 mechanism could proceed through rearrangements to form a more stable carbocation intermediate. Since an E1 mechanism has the same first step as an SN1 mechanism, would E1 reactions be likely to undergo rearrangements?
(*Circle one*) yes / no.

(b) Draw the mechanism to rationalize the formation of the following rearranged product.

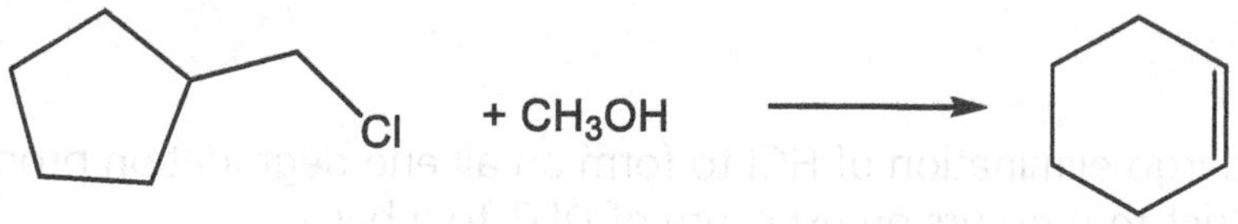

Reflection: on a separate sheet of paper.
As a group, describe three concepts your group has learned from this activity and the one most important unanswered question about this activity that remains with your group. Turn this in before leaving class.

Additional Questions:

16. For the following reactions, identify all the β-hydrogens, draw all possible elimination products, and circle the major elimination product formed.

(a)

(b)

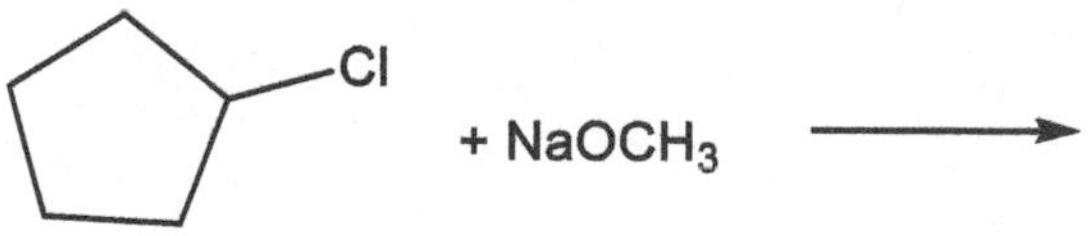

(c)

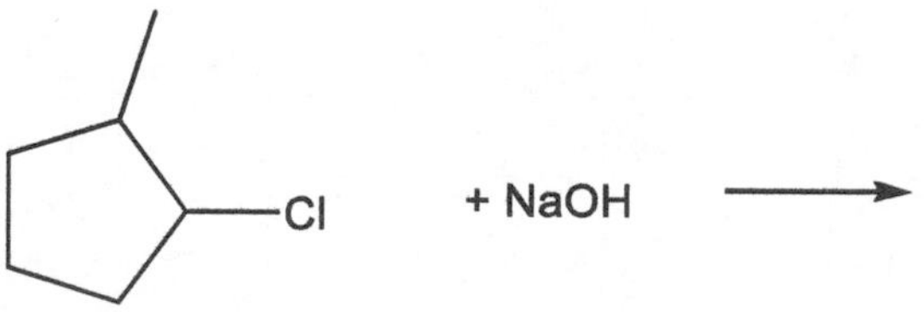

(d)

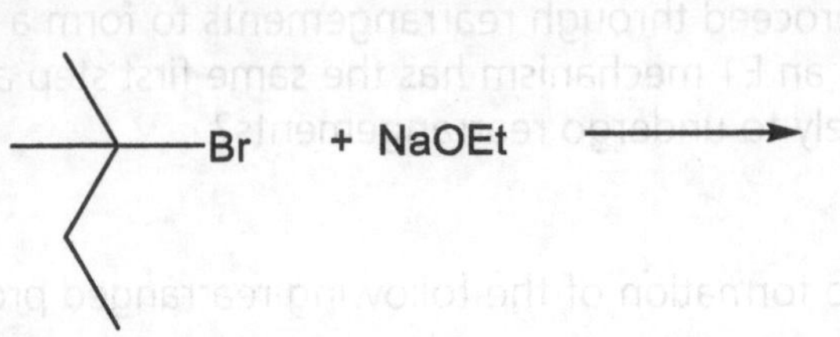

17. The pesticide DDT can undergo elimination of HCl to form an alkene degradation product. Draw the elimination product that occurs on exposure of DDT to a base.

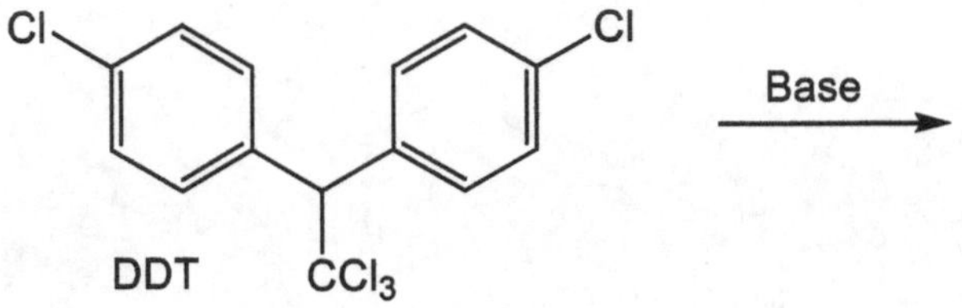

Base ⟶

Class Activity 10B

Stereochemistry of E2 Elimination

Prior Knowledge:
Before beginning this activity, students should be familiar with the following concepts:

- Functional groups, naming alkenes, E/Z and R/S nomenclature.
- Newman projections
- Curved arrows.
- β–elimination reactions, Greek characters
- Substitution reactions, SN1 and SN2 mechanism (only for Q#14 and Q#15).

Learning Objectives
Content Learning Objectives:
After completing this activity students should be able to:
- Interconvert Newman projections to orient groups ANTI.
- Determine alkene geometry of a β–elimination reaction, based on ANTI elimination.
- Predict and draw the major elimination product for acyclic and cyclic compounds.

Process Objectives:
- Information Processing. Students combine and transform information from elimination reactions, stereochemistry and Newman projections.
- Critical Thinking. Students evaluate and synthesize information about stereochemistry to reach a conclusion about the elimination products.

Class Activity 10B

Stereochemistry of E2 Elimination

Model 1: Stereochemistry of E2 Elimination
In any E2 elimination reaction the two groups that are eliminated must be anti (180° apart).

Ph = phenyl = C_6H_5=

Questions:

1. For the reaction shown in Model 1:
   (a) In structure A, label the α carbon and any β carbons.
   (b) When converting the wedge dash structure A to the Newman structure B, which carbon is in the front? (*Circle one*) α / β. Confirm that A and B are the same conformation.
   (c) To convert B to C, groups on the (*circle one*) α / β carbon were rotated. Confirm that C and D are the same conformation.
   (d) Elimination of the β-H only occurs when the H and Br are ANTI or 180° apart. In which Newman projection is the H and Br anti? (*Circle one*) B / C.
   (e) In structure C, the Ph and CH_3 groups on the (*circle one*) same / opposite sides.
   (f) In the product alkene E, the Ph and CH_3 groups on the (*circle one*) same / opposite sides. Confirm that the Newman E and the alkene are the same compound.

2. The three possible staggered conformations of (S)-2-bromobutane are shown below.
   (a) Cross out the <u>one</u> conformation that cannot undergo an E2 elimination from the conformation shown (with no further rotation).

(b)   Circle the one of the remaining two conformations that is more stable, or of lower potential energy. Explain.

(c)   From the <u>two</u> conformations that can undergo E2 elimination, use curved arrows to show the flow of electrons during each E2 reaction, and draw the alkenes that will form upon elimination.

(d)   Determine whether the two alkene products are E or Z geometry. Are the two alkenes the same or isomers?

(e)   Once your group has reached agreement on the above questions, predict which alkene would be the major product (Hint; keep in mind your answer for #2b). Explain.

3. (a)   Draw a wedge dash and a Newman representation (as viewed along the C3C4 bond) of (S)-3-bromohexane.

(b)   Rotate the Newman projection so the bromine and the β–hydrogen (at C3 and C4) are anti. There are two conformations where Br and H are anti; draw both conformations and determine which is the most stable. (Refer to #2a).

(c)   Draw the product that would result from β–elimination along the C3C4 bond from the most stable conformation (clearly show geometry of alkene).

(d)   Provide a name for the alkene formed above.

Model 2: Elimination in Cyclic Compounds

Questions:

4. For the compound shown in Model 2:
    (a)   Label the β–carbons. Circle any β–H.

    (b)   Draw the <u>other</u> chair conformation of the cyclohexane shown. Label the β–H.

    (c)   In either conformation are the Br and the β–H ANTI to each other?
            (*Circle one*) yes / no.

    (d)   Predict whether β–elimination would occur from the structure shown in Model 2. Explain.

5. The structure in Model 2 has the Br and CH_3 group trans. Draw another structure where the Br and CH_3 are cis. Draw both chair conformations of this compound. Once your group has reached consensus, determine if either chair conformation is capable of undergoing β–elimination. Explain.

6. For the following compounds:
    (a)   (b)

(i)   Label all the β–hydrogens.
(ii)   Determine which hydrogens will undergo elimination by redrawing the compound (by rotating single bonds only) to see if the H and X can become ANTI. If elimination can occur, draw the product obtained.
(iii) If more than one product can be formed, predict the major isomer.

Reflection: on a separate sheet of paper.
As a group, describe three concepts your group has learned from this activity and the one most important unanswered question about this activity that remains with your group. Turn this in before leaving class.

Additional Questions:
7. For the compounds shown below, identify the β–hydrogens, determine if the H can be rotated anti to the leaving group and draw the elimination products expected. Circle the major product.

(a)

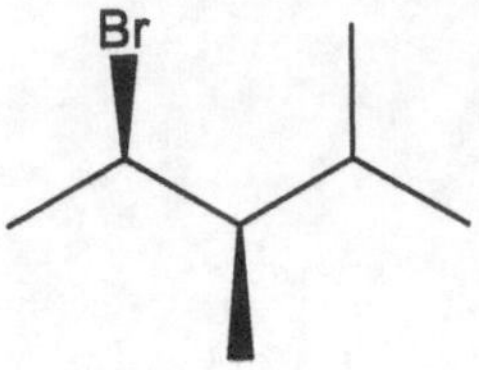

(b)

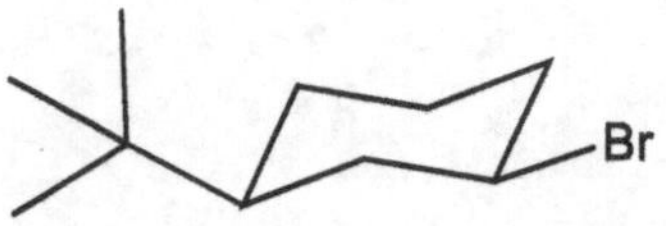

(c)

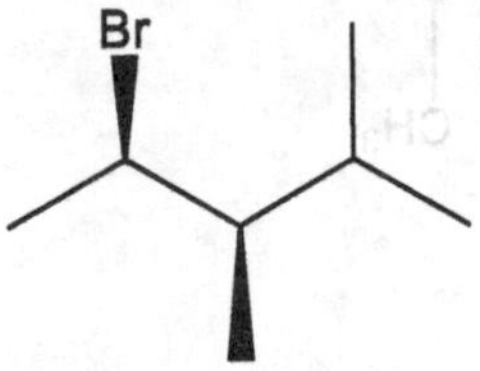

(d)

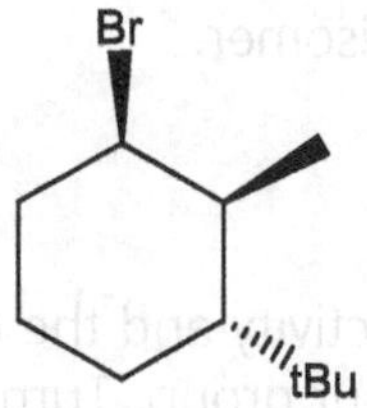

8. For the following reactions, identify RX type (1°, 2°, 3°), solvent type (protic or aprotic), and nucleophile/base strength (strong or weak). Based on the above, predict the mechanism (SN1, SN2, E1, E2) and draw the major product.

(a)

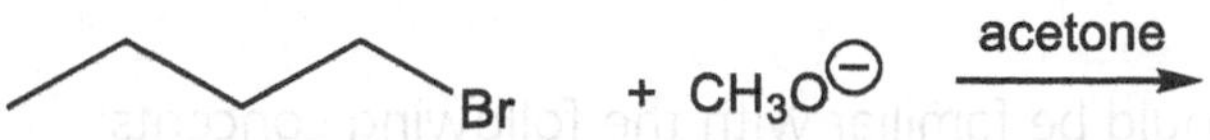

+ $CH_3O^\ominus$ $\xrightarrow{\text{acetone}}$

(b)

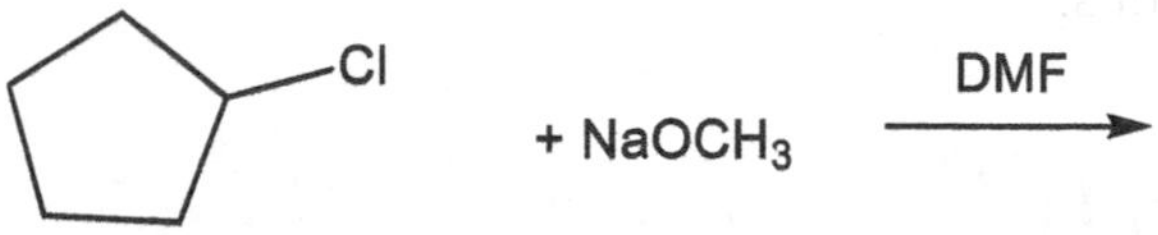

+ $NaOCH_3$ $\xrightarrow{\text{DMF}}$

(c)

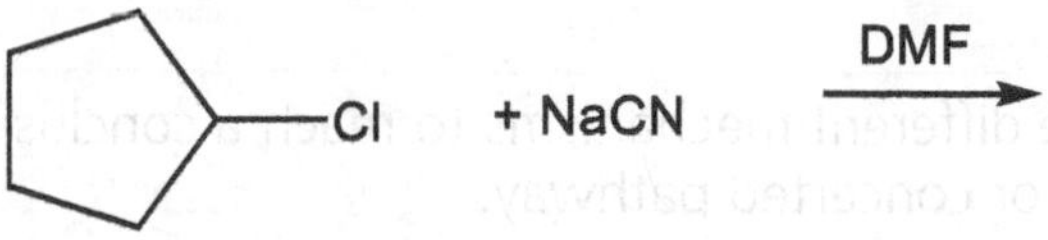

+ NaCN $\xrightarrow{\text{DMF}}$

(d)

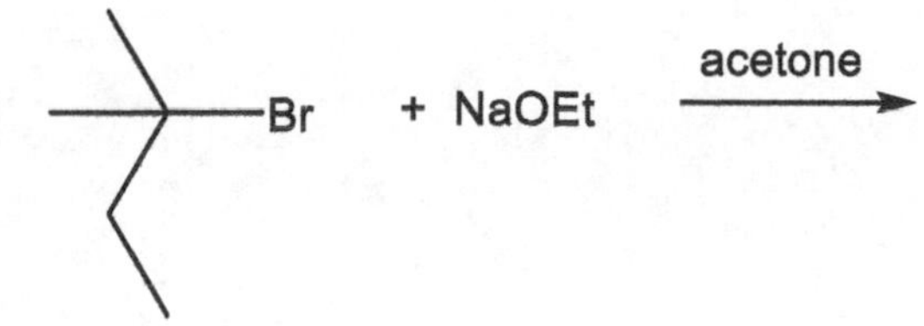

+ NaOEt $\xrightarrow{\text{acetone}}$

(e)

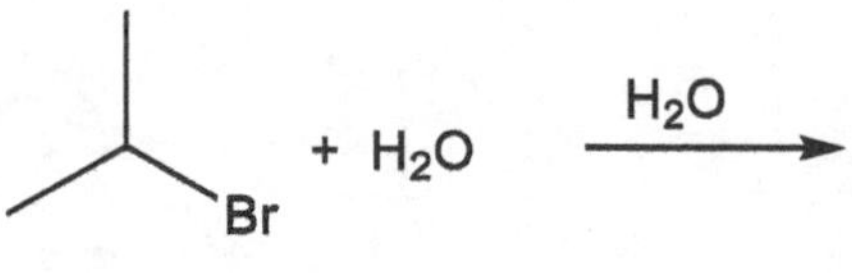

+ H_2O $\xrightarrow{H_2O}$

Class Activity 11A

Electrophilic Addition

Prior Knowledge:
Before beginning this activity, students should be familiar with the following concepts:

- Functional groups, naming alkenes, E/Z nomenclature.
- Curved arrows.
- Cationic intermediates (classification as primary etc., stability).
- Nucleophile and electrophile definitions.

Learning Objectives
Content Learning Objectives:
After completing this activity students should be able to:
- Identify an addition reaction.
- Determine that addition reactions occur through a stepwise mechanism.
- Predict and draw the major addition product from addition to unsymmetrical alkenes.

Process Objectives:
- Critical Thinking. Students evaluate the different mechanisms to reach a conclusion about whether addition occurs by a stepwise or concerted pathway.
- Problem Solving. Students identify the intermediates and plan a strategy that will lead to the products based on determining the stability of the cationic intermediate.

Class Activity 11A

Electrophilic Addition

Model 1: Electrophilic Addition

$$\text{(C}_6\text{H}_{12}\text{)} \quad + \quad Cl-\!\!-Cl \quad \longrightarrow \quad Cl-\!\!<\!\!>\!\!-Cl \quad \text{(C}_6\text{H}_{12}\text{Cl}_2\text{)}$$

Questions:

1. Consider the reaction shown in Model 1:

   (a) How many compounds are present on the reactant side of the equation? _____

   (b) How many compounds are present on the product side? _____

2. Describe any changes that occur in the reaction shown in Model 1.

3. What reaction type best describes the net reaction illustrated in Model 1? (*Circle one*)
   substitution / elimination / addition.
   Explain your choice.

4. Classify the following reactions as substitution, elimination or addition reactions.

   (a)

   $$\text{acetone} \quad + \; H_2 \quad \longrightarrow \quad \text{isopropanol}$$

   (b)

   $$\text{cyclopentyl}-Br \; + \; NaCN \quad \longrightarrow \quad \text{cyclopentyl}-CN \; + \; NaBr$$

(c)

(d)

Model 2: Possible Mechanisms of Addition Reactions

Two mechanisms can be proposed for the addition reaction to an alkene.

Questions:

5. Consider the <u>stepwise</u> mechanism show in Model 2:

   (a) How many steps are in the stepwise mechanism? _____

   (b) What role does the alkene play when it reacts with HCl in step 1?
        (*Circle one*) acid / base / nucleophile / electrophile.

   (c) What intermediate is formed after step 1? (*Circle one*) cation / anion / radical.

   (d) What role does the intermediate play when it reacts with Cl⊖ in step 2?
        (*Circle one*) acid / base / nucleophile / electrophile.

   (e) Curved arrows are illustrated for the stepwise mechanism. What do the curved arrows represent?

6. Consider the <u>concerted</u> mechanism show in Model 2:
   (a) How many steps are in the concerted mechanism? _____

   (b) Is the intermediate cation formed in the concerted mechanism? (*Circle one*) yes / no.

   (c) Compare the products for the concerted and stepwise mechanism. The products for both mechanisms are the (*circle one*) same / different.

   (d) Draw curved arrows to illustrate electron movement for the concerted mechanism. (HINT: recall that an arrow shows electron movement for every bond broken and formed.)

7. As a group discuss whether the mechanism for addition to an alkene would be concerted, or stepwise, or if there is not enough information to determine the mechanism. Once you have reached consensus, write a statement explaining your choice of mechanism.

Model 3: Addition of HCl

Questions:
8. (a) In Eq. 1 of Model 3, the alkene is reacted with HCl only. What product is formed?

   (b) In Eq. 2 of Model 3 the alkene is reacted with HCl and NaBr. What product(s) is/are formed?

   (c) Describe what is different between Eq. 1 and Eq. 2.

9. In your groups, discuss whether the outcome from Eq. 1 helps determine which mechanism (stepwise or concerted) from Model 2 is more likely. Explain.

10. Consider the two products formed in the reaction shown in Eq. 2. Discuss whether the stepwise or the concerted mechanism best explains the formation of these two products. Once consensus has been obtained, explain your mechanistic choice.

11. In determining how to draw a reaction mechanism, it is helpful to identify the roles of the reagents and to proceed one step at a time, using curved arrows to show the electron movement for every bond broken and formed. In the reaction of an alkene with HCl:

    (a) HCl acts as the (*circle one*) acid / base, and will (*circle one*) donate / accept an electron pair.

    (b) The alkene acts as the (*circle one*) acid / base, and will (*circle one*) donate / accept an electron pair.

    (c) The first step of the stepwise addition of H⊕ to an alkene is shown below. Label the reagents as acid and base and draw curved arrows to indicate the electron movement.

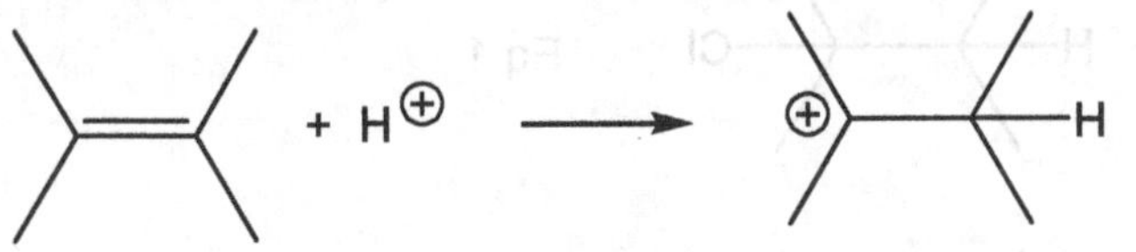

12. Draw the complete stepwise mechanism, using curved arrows, to explain the formation of the two products formed in Eq. 2, shown below. (HINT: use #11c to start the mechanism)

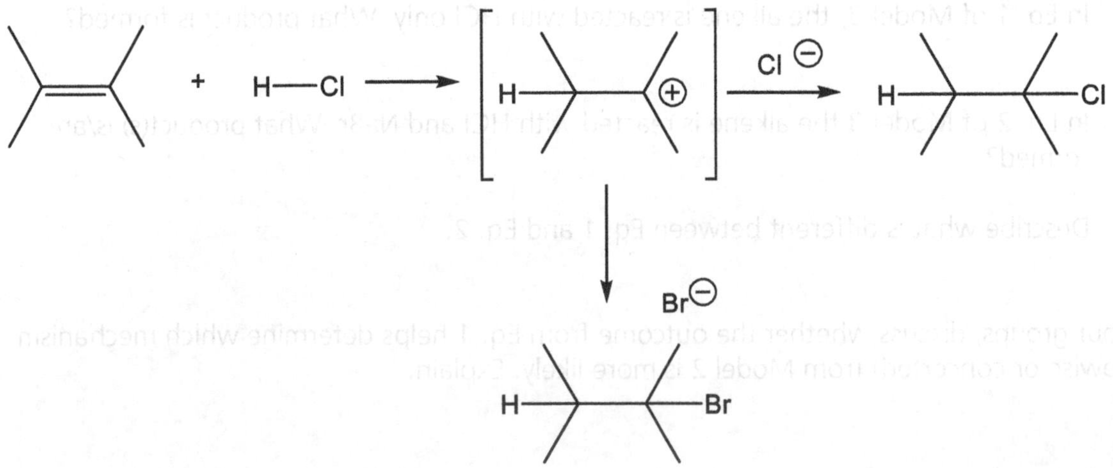

Model 4: Addition to Unsymmetrical Alkenes

Questions:

13. (a) In structure A from Model 4, compare the two carbons of the alkene (C2 and C3). How many R groups are on C2? _____ How many R groups are on C3? _____

    (b) Structure A is considered to be an unsymmetrical alkene. Explain why.

    (c) When HCl is added to compound A, two products are possible, B and C. For product B, which carbon did the H from HCl add to? (*Circle one*) C2 / C3. For product C, which carbon did the H add to? (*Circle one*) C2 / C3.

14. (a) Given that a stepwise mechanism of addition takes place, draw the carbocation intermediate (in the box above) that would result in product B and the intermediate that would result in product C. Label each cation as 1°, 2°, or 3°.

    (b) Which cation would be the most likely to be formed? Explain.

    (c) Only one of these products (B or C) forms. Based on your answer to #14b above, put an X through the product that does not form in Model 4.

    (d) Once your group has agreed which product is most likely to form (B or C), draw curved arrows in Model 4 to show electron movement to form this product.

15. (a) Markovnikov's Rule states in an electrophilic addition reaction the hydrogen adds to the carbon of the alkene that has the most hydrogens. Which product from Model 4 is the Markovnikov product? (*Circle one*) B / C. Write 'Markovnikov' next to the appropriate product in Model 4.

(b)   If the electrophile is <u>not</u> hydrogen, would you expect Markovnikov's rule to be followed, where the electrophile adds to the carbon with the most hydrogens? Explain.

(c)   Once your group agrees on the above questions, rewrite Markovnikov's rule using terms such as electrophile and carbocation.

Reflection: on a separate sheet of paper.
As a group, describe three concepts your group has learned from this activity and the one most important unanswered question about this activity that remains with your group. Turn this in before leaving class.

Additional Questions:

16. (a)   Use curved arrows to illustrate electron movement, and draw the carbocation intermediate <u>first</u> formed from the reaction of HBr and the alkene shown below.

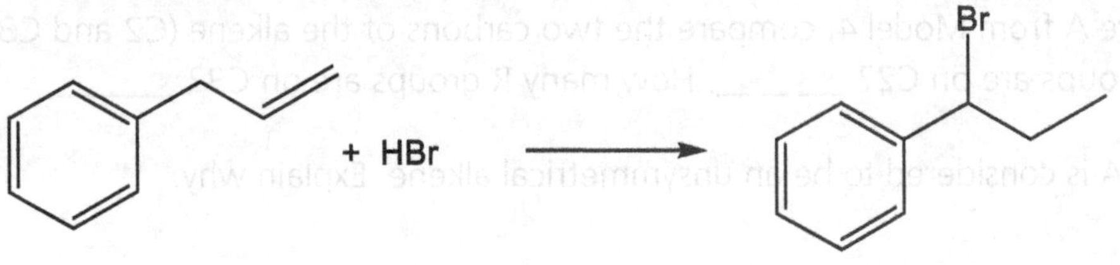

(b)   Would the carbocation intermediate first formed in #16a, lead to the product shown above? Explain.

(c)   Draw the carbocation intermediate that would lead directly to the product shown above.

(d)   Which carbocation, from #16a or #16c, is the more stable cation? Explain your answer by drawing resonance forms if possible.

(e)   Justify using curved arrows to show electron movement, how the initially formed cation (from #16a) can be converted to the cation leading to the product (from #16c). Draw in all hydrogens to help explain your answer.

17. Would an addition reaction occur faster in a polar solvent or a non-polar solvent? (Refer to the mechanism in Model 2). Explain your answer.

18. Formation of polymers, large molecules composed of many smaller units called monomers, can occur via addition of many alkenes together to form a long chain. Predict a mechanism for formation of polyethylene.

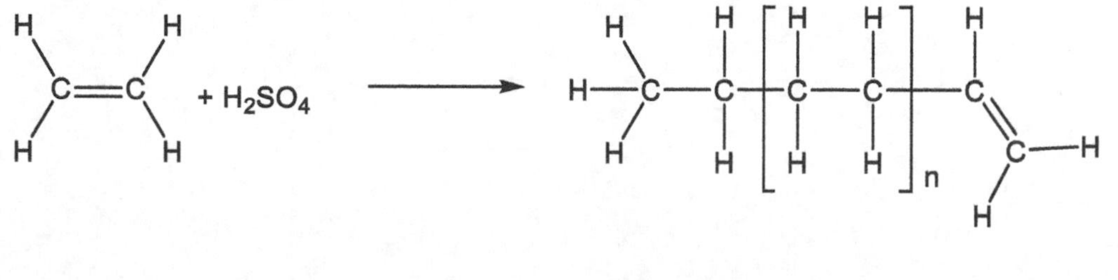

Class Activity 11B

Electrophilic Addition
Part B: Addition of Oxygen

Prior Knowledge:
Before beginning this activity, students should be familiar with the following concepts:

- Lewis acid/base definition.
- Addition of HX to an alkene.
- Markovnikov's rule, regiochemistry of addition.
- Curved arrows.
- Cationic intermediates (classification as primary etc., stability).
- Nucleophile and electrophile definitions.

Learning Objectives
Content Learning Objectives:
After completing this activity students should be able to:
- Predict the products of hydration reactions, and determine the correct regiochemistry and stereochemistry.
- Draw the mechanism for acid catalyzed hydration of an alkene.
- Draw intermediates that form in oxymercuration and hydroboration and predict partial mechanisms of these reactions.

Process Objectives:
- Information Processing. Students explore and interpret three types of hydration reactions.
- Critical Thinking. Students analyze these reactions and mechanisms to determine why there are particular regiochemical and stereochemical outcomes.

Class Activity 11B

Electrophilic Addition
Part B: Addition of Oxygen

Model 1: Hydration of Alkenes

The addition of water to an alkene (hydration) can be accomplished using several different reagents. These methods may vary in the regio- and stereochemistry of the products.

H_2SO_4
H_2O

1. $Hg(OAc)_2$, H_2O
2. $NaBH_4$

1. BH_3
2. H_2O_2, $HO^\ominus$

OH

OH

H   OH

Questions:

1. (a) Compare the products of the three hydration reactions shown in Model 1. What functional group is formed on hydration of an alkene? _____

   (b) Which reactions in Model 1 afford products that have the same regiochemistry of addition? (i.e. addition occurred to give the same constitutional isomer).

2. For each reaction determine if the H adds to the least or most substituted carbon of the alkene. Label each product from Model 1 as Markovnikov or anti-Markovnikov.

3. Indicate whether each reaction in Model 1 occurs to give SYN addition, ANTI addition or if it cannot be determined from the scheme above.

Model 2: Acid Catalyzed Hydration

Questions:

4. (a) In the reaction shown in Model 2, what reagents are reacted with the alkene?

   (b) H⊕ acts as the (*circle one*) acid / base.
   (c) Recall the Lewis acid/base definition. An acid will (*circle one*) donate / accept electrons.
   (d) The alkene will most likely (*circle one*) donate / accept electrons

5. (a) Consider the mechanism for addition of HX as shown below (from class activity 11A). Label each reagent as nucleophile, electrophile, acid or base for both steps.

   (b) From the mechanism above, what is the <u>first</u> thing that happens when an alkene is combined with an acid?

   (c) The curved arrow showing the formation of the C-H bond "bounces" off one of the carbons of the double bond. Explain what this means.

6. (a) The mechanism for hydration is similar to the mechanism for HCl addition shown in #5a above. Draw the cation obtained on reaction of the alkene shown below with acid, using curved arrows to show electron movement.

(b)   The proton should add to the (*circle one*) most / least substituted carbon of the alkene. Explain your reasoning.

(c)   In the second step, the cation will act as a (*circle one*) Nu⊖ / E⊕ .

(d)   Consider the reagents present in the acid catalyzed hydration (alkene, H⊕ and H_2O). Which reagent is most likely to react with the intermediate cation?_____ This reagent would most likely act as a (*circle one*) Nu⊖ / E⊕.

(e)   Draw the mechanism for the second step above, using curved arrows. Remember that the reaction is conducted in water (not hydroxide). Is the product of the second step the same as the final product? (*Circle one*) yes / no.

If not what remaining step must occur to obtain the final product?

7. Once your group has reached consensus, discuss what the product might be in a reaction of the alkene shown in Model 2 with H⊕ and CH_3OH. Draw a mechanism for this reaction and explain how it differs from the mechanism for the reaction conducted in H_2O.

Model 3: Oxymercuration/Demercuration

Questions:

8. The reagent Hg(OAc)₂ can be viewed as dissociated into ⊕HgOAc and ⊖OAc. Of these two species, which would most likely add to the alkene? (Compare to step 1 of acid catalyzed reaction).

9. (a)   Instead of forming a carbocation, a more stable cyclic mercurinium ion is formed. If one of the C-Hg bonds of the three membered ring started to break, the carbon more likely to hold the partial positive charge would be the (*circle one*) most / least substituted carbon. Place the δ+ sign on the carbon you chose.

    (b)   The mercurium ion intermediate will react with one of two nucleophiles present in the reaction, ⊖OAc or H₂O. If the solvent, water, is present in great excess, which reagent will act as the nucleophile? (*Circle one*) ⊖OAc / H₂O.

          Is the reagent circled the better nucleophile? _____ If not explain why it was chosen as the Nu⊖.

10. The last step in Model 3 is called demercuration. What bond is being broken and what bond is being formed on the molecule shown?

11. Once your group has reached agreement on the above questions, discuss what product would result if the reaction in Model 3 was conducted in CH₃OH as a solvent instead of H₂O.

Model 4: Hydroboration

$$H_3C \diagdown \diagup H \quad + \quad H-B \text{(with H's)} \quad \longrightarrow \quad H_3C - \overset{H}{\underset{H_3C}{|}} - \overset{BH_2}{\underset{H}{|}} - H$$

Questions:

12. (a)   How many valence electrons does boron have in the compound BH₃ ? _____

    (b). The electronegativity value for H = 2.3 and for B = 2.1.
          Which is more electronegative? (*Circle one*) B / H.

    (c)   Draw arrows on the structure for BH₃ above to show the dipole of the B-H bond.

    (d)   Which <u>atom</u> of BH₃ is more likely to act as the electrophile? (*Circle one*) B / H.

13. In the mechanism for hydroboration, the B and H add to the alkene at the same time (<u>concerted</u> , shown below). Although a full carbocation is not formed, the transition state shows that a partial positive charge develops on the carbon that is not bonded to the electrophile, boron.

(a) Why does the addition occur to give the regiochemistry shown above? (Consider the stabillity of partial positive charge.)

(b) According to this mechanism, does addition occur SYN or ANTI?

(c) Hydroboration is <u>stereospecific</u> because during the first step the boron and hydrogen are added at the same time to the double bond. Would you expect addition to occur from the top face, the bottom face, or both faces equally?

14. In the oxidation step of the hydroboration/oxidation sequence, the boron is oxidized to OH with hydrogen peroxide and base. During oxidation, the stereo- and regiochemistry is not changed.

(a) Label the products as Markovnikov or anti-Markovnikov.

(b) Would you expect the products above to be formed in different or equal amounts. Explain.

(c)   Compare the hydration of an alkene using hydroboration vs acid in water.

15. In your groups compare the three different methods for hydration of an alkene. Discuss reasons why one method might be preferred over another method.

Reflection: on a separate sheet of paper.
As a group, describe three concepts your group has learned from this activity and the one most important unanswered question about this activity that remains with your group. Turn this in before leaving class.

Additional Questions:
16. Indicate whether the product in each of the following reactions is Markovnikov or anti-Markovnikov. Determine the reagents that will result in the product given (if possible).
(a)

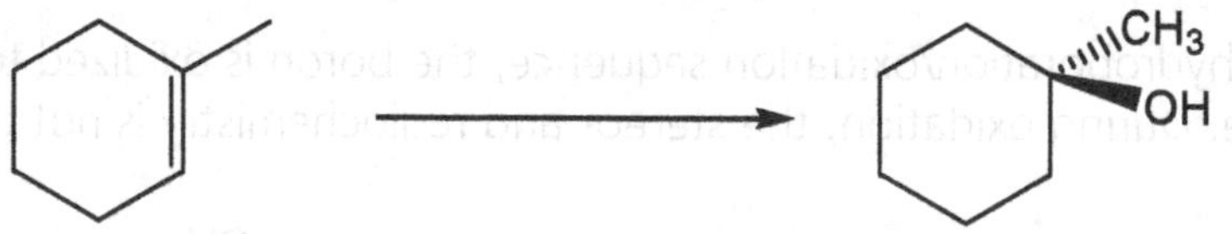

(b)

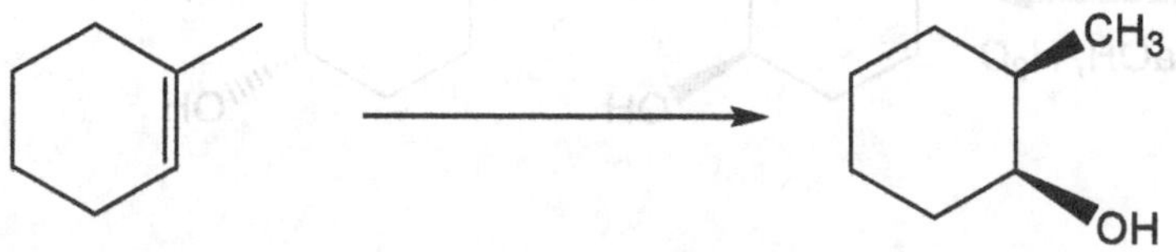

(c)

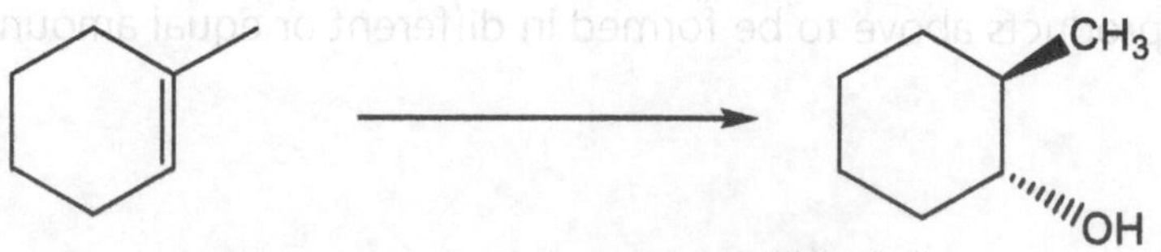

(d)

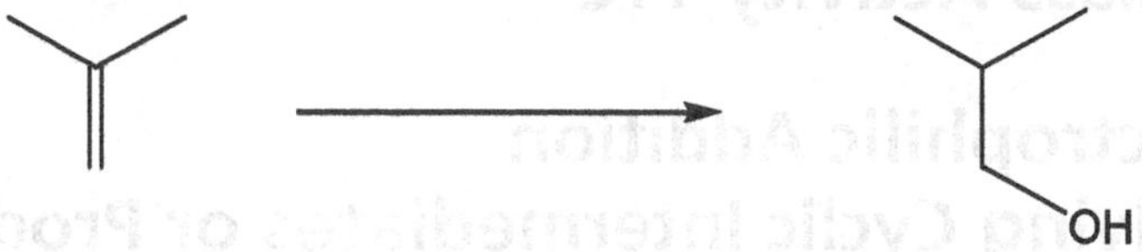

(e)

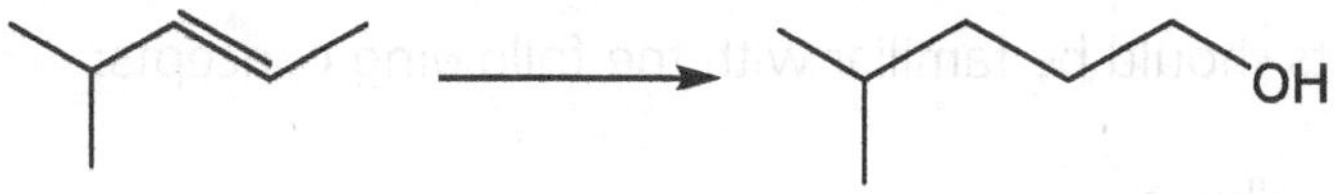

(f)

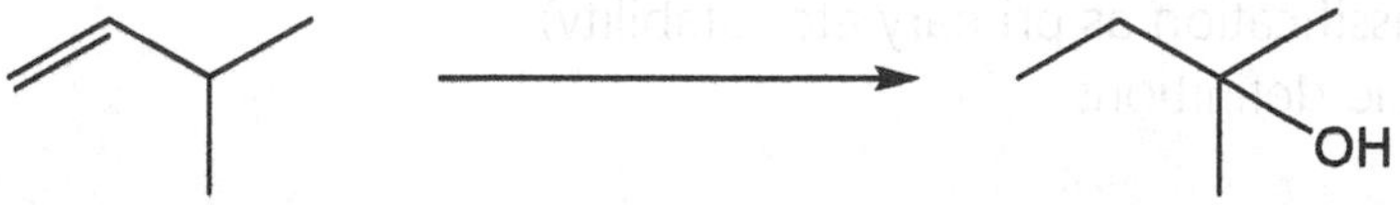

17. Draw the complete mechanism for the following reactions, using curved arrows to show electron movement.

(a)

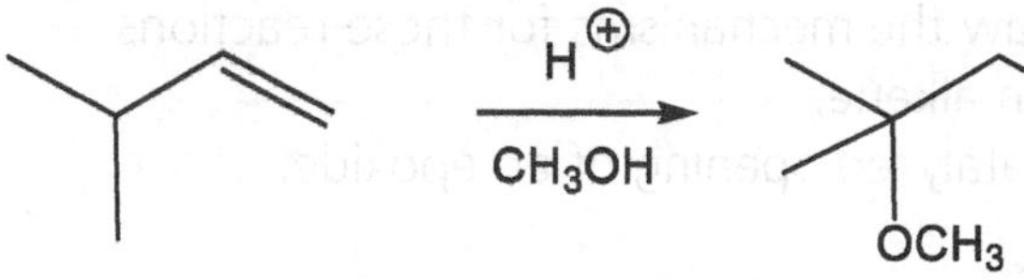

(b)

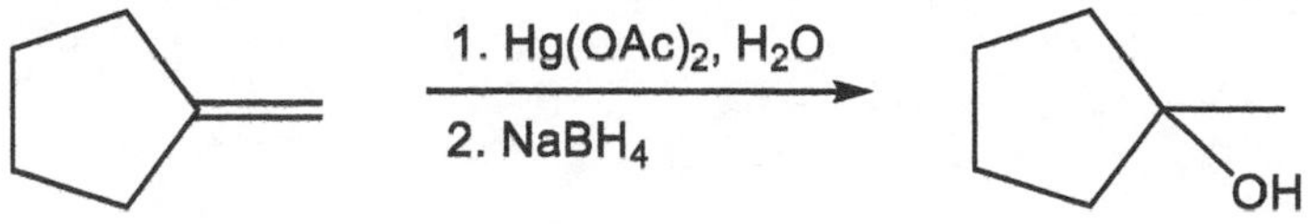

Class Activity 11C

Electrophilic Addition
Part C: Additions Involving Cyclic Intermediates or Products

Prior Knowledge:
Before beginning this activity, students should be familiar with the following concepts:

- Addition of HX and H_2O to an alkene.
- Markovnikov's rule, regiochemistry of addition.
- Curved arrows.
- Cationic intermediates (classification as primary etc., stability).
- Nucleophile and electrophile definitions.

Learning Objectives
Content Learning Objectives:
After completing this activity students should be able to:
- Predict the products with correct regiochemistry and stereochemistry of bromination and halohydrin formation reactions, and draw the mechanisms for these reactions.
- Predict the product of epoxidation of an alkene.
- Draw the products expected for base catalyzed opening of an epoxide.

Process Objectives:
- Problem Solving. Students identify variables in reactions in order to make conclusions about the reaction mechanisms and products.
- Information Processing. Students explore and manipulate information from a variety of reactions and concepts

Class Activity 11C

Electrophilic Addition
Part C: Additions Involving Cyclic Intermediates or Products

Model 1: Halogenation of Alkenes

cyclic bromonium ion   vicinal dibromide

Questions:

1. (a) In the halogenation reaction shown in Model 1, what reagent is added to the alkene?

   (b) What is the functional group of the product that is formed on halogenation?

   (c) Describe what the term 'vicinal' is refers to.

   (d) Compare the Br atoms in the product. Halogenation occurs (*circle one*) SYN / ANTI.

2. (a) In general, an alkene is (*circle one*) electron rich / electron deficient. The alkene would act as a (*circle one*) Nu⊖ / E⊕.

   (b) If an electron rich species approaches one end of the Br-Br bond, discuss what affect this would this have on the dipole of the Br-Br bond. Draw δ+ and δ- signs on the bromine in Model 1 to illustrate this induced dipole. Label the <u>atom</u> that would act as the electrophile.

3. (a) In step 1, identify what bonds are being broken and formed.
   Bonds broken:

   Bonds formed:

   (b) Draw curved arrows for step 1 to illustrate electron movement. Where must the electrons come from to form the second C-Br bond in the cyclic bromonium ion?

(c)   The cyclic bromonium ion results from forming the second C-Br bond. If this bond did not form, the cation below would have resulted. Compare the stability of this cation to the cyclic bromonium ion. Which would you expect to be more stable? Explain.

4. (a)   For step 2, label the Nu⊖ and E⊕, and draw curved arrows to show electron movement.

    (b)   The Br⊖ attacks from the (*circle one*) same / opposite side as the Br in the cyclic bromonium ion ring. Explain why you expect attack from the face chosen.

5.   When bromination is conducted in H_2O, a halohydrin is formed as shown below.

+ Br—Br ⟶ H_2O ⟶

halohydrin A
(only product)   halohydrin B
(not formed)

    (a)   What two groups make up a halohydrin? _____ Compare the stereochemistry of these groups. The addition occurred (*circle one*) SYN / ANTI.

    (b)   The cyclic bromonium ion is also the intermediate in halohydrin formation. What reagent acts as the nucleophile that adds to the cyclic bromonium ion?

          Use curved arrows to show the nucleophile attacking the cyclic bromonium ion.

    (c)   The two possible halohydrin products, A and B, are shown. What is the relationship between these compounds? (*Circle one*) conformational / constitutional / stereoisomers.

    (d)   Halohydrin A is the only product formed. In order to form halohydrin A, the water attacked the (*circle one*) more / less substituted carbon.

(e) Since only product A is formed, this reaction is said to be <u>regioselective</u>. As a group discuss why attack is regioselective. (HINT: consider which one of the C-Br bonds of the cyclic bromonium ion would be more likely to break).

Is the regioselectivity in this reaction controlled by electronic factors or by steric factors?

Is the mechanism S_N1-like or S_N2-like? Explain.

Model 2: Epoxidation of Alkenes

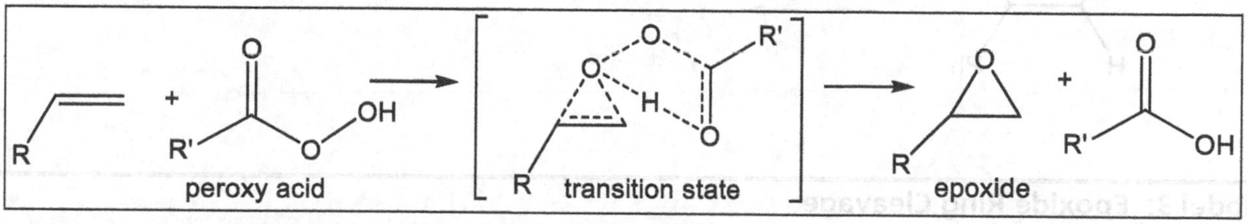

Questions:

6. (a) In an epoxidation, what <u>atom</u> is added to the alkene to form the epoxide?

(b) What reagent in Model 2 reacts with the alkene to form the epoxide?

(c) The O-O bond of the peroxy acid is weak, and can break heterolytically as shown below. Draw the ions obtained for each cleavage depicted by the arrows shown below.

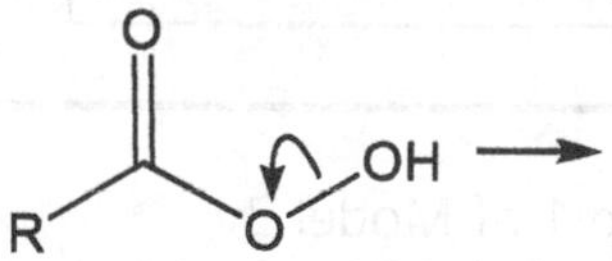

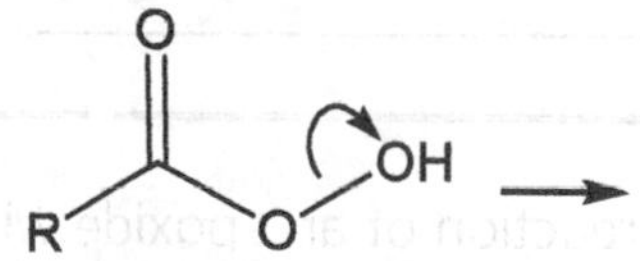

(d) Compare the two anions above. Which would be the better leaving group? Explain.

(e) Based on your answer above, predict which oxygen of the peroxyacid is likely to act as the electrophile. Circle the oxygen in the peroxy acid in Model 2.

7. (a) Epoxidation occurs by forming the transition state shown in Model 2. The mechanism is (*circle one*) stepwise / concerted.

(b)   Does the mechanism support SYN or ANTI addition? Explain.

8. The geometry of the alkene is <u>retained</u> in epoxidation reactions. This means that if two groups
    on the alkene were cis, they would remain cis in the epoxide product.
    (a)   In the following scheme, circle the products that clearly show <u>retention</u> of alkene
          geometry on epoxidation with a peroxyacid.

    (b)   Draw the products formed on reaction of the following alkene with a peroxyacid,
          showing syn addition and retention of alkene geometry.

Model 3: Epoxide Ring Cleavage

Questions:
9. Consider the reaction of an epoxide with acid as shown in Step 1 of Model 3
    (a)   The C-O bonds of an epoxide are (*circle one*) polar / non-polar. If polar label each C-O
          bond with δ+ and δ– to indicate the polarity.

    (b)   An acid donates a proton. What atom of the epoxide will be protonated? _____ Draw
          curved arrows on the structure above to illustrate formation of the protonated epoxide.
    (c)   The protonated epoxide is (*circle one*) more / less reactive than the neutral epoxide.

10. In step 2, the protonated epoxide forms a 1,2-diol.
    (a)   What compound acts as the nucleophile in step 2?
    (b)   The product 1,2-diol is (*circle one*) cis / trans.

(c)   Compare the second step of the epoxide ring opening to the second step of halogenation of alkenes (Model 1). Note any similarities and differences in the reactions.

(d)   Predict the product formed if the reaction is conducted in CH_3OH instead of H_2O.

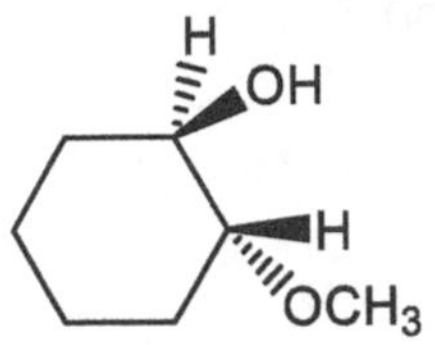

11. (a)   If the epoxide is NOT activated by reacting with acid, then any nucleophiles that react with the neutral epoxide must be is (circle one) strong / weak.

(b)   What atom of the epoxide would most likely be attacked by a nucleophile? Explain why.

(c)   Draw curved arrows to show how the nucleophile attacks the epoxide in the reaction below. As a group, discuss why the nucleophile attacks on the opposite side of the epoxide oxygen.

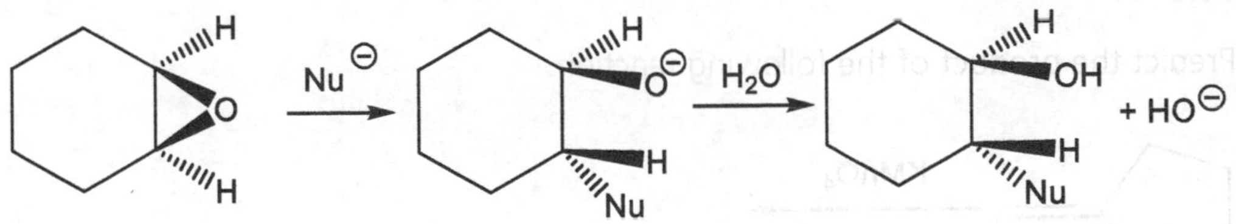

(d)   Consider the unsymmetrical epoxide shown below. On reaction with a nucleophile, which carbon would most likely be attacked, the more or less substituted? (HINT: since the epoxide is not protonated, a partial cation is not likely to be formed). Is the regioselectivity in this reaction controlled by electronic factors or by steric factors? Draw the product expected.

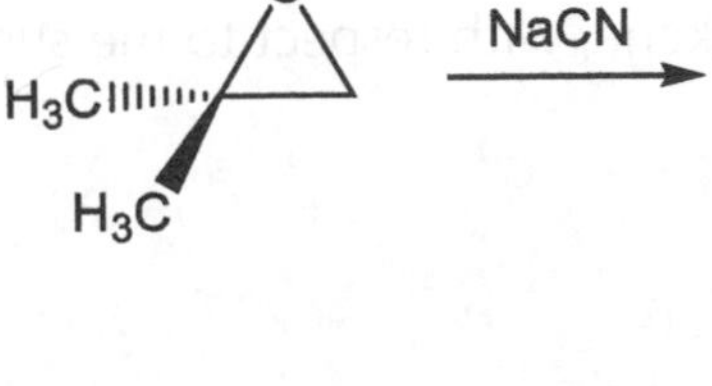

Model 4: Oxidation and Oxidative Cleavage

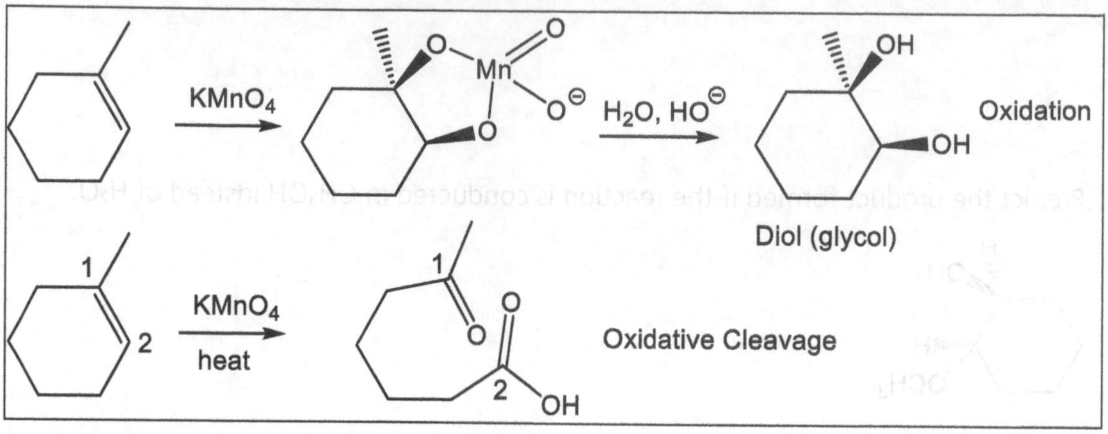

Diol (glycol)

Oxidation

Oxidative Cleavage

Questions:

12. For the <u>oxidation</u> reaction shown in Model 4.
   (a) What product is formed?

   (b) The stereochemistry of the product is (*circle one*) cis / trans.

   (c) Consider the cyclic intermediate. How does the stereochemistry of the intermediate dictate the stereochemistry of the product?

   (d) Oxidation can also occur using OsO_4 and H_2O_2 to afford the same product. Would you expect a similar cyclic intermediate to be formed as well? (*Circle one*) yes / no.

   (e) Predict the product of the following reaction.

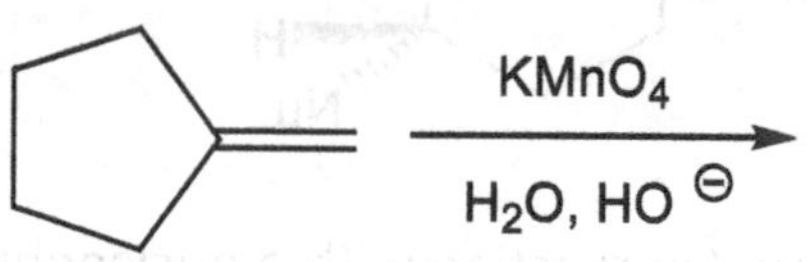

13. For the <u>oxidative cleavage</u> reaction shown in Model 4.
   (a) What reaction conditions have been changed in order to give the cleavage product?

   (b) Carbon 1 of the alkene formed a (*circle one*) ketone / aldehyde / carboxylic acid.
   (c) Carbon 2 of the alkene formed a (*circle one*) ketone / aldehyde / carboxylic acid.
   (d) What is the difference between C1 and C2 of the alkene, with respect to the substitution of the alkene (i.e. mono, di, tri, tetra).

(e)   As a group, determine a rule for how differently substituted alkenes are oxidized.

(f)   Predict the products of the following reaction.

$$\text{(structure)} \xrightarrow[\text{heat}]{\text{KMnO}_4}$$

14. Oxidative cleavage can also occur in the presence of ozone, O_3, as shown below.

$$\text{(structure)} \xrightarrow[\text{(CH}_3)_2\text{S}]{O_3} \text{(structure)}$$

(a)   Compare the product of the ozone reaction above to the oxidative cleavage product shown in Model 4. What is different (note reaction at C1 and C2)?

(b)   Determine a rule for how differently substituted alkenes are oxidized with ozone.

(c)   Predict the products of the following reaction.

$$\text{(structure)} \xrightarrow[\text{(CH}_3)_2\text{S}]{O_3}$$

Reflection: on a separate sheet of paper.
   As a group, describe three concepts your group has learned from this activity and the one most important unanswered question about this activity that remains with your group. Turn this in before leaving class.

Additional Questions:
15. Draw the missing reagents/products in the following reactions.

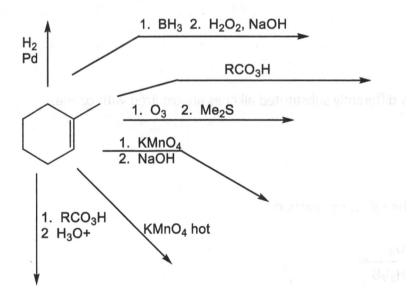

16. Draw the products of the following reactions. Show stereochemistry if necessary.

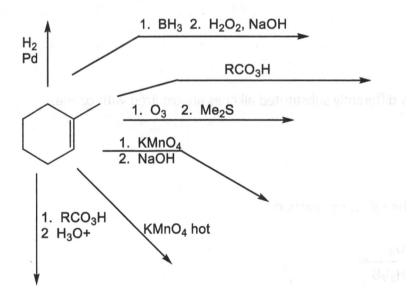

Class Activity 12

Electrophilic Addition to Alkynes

Prior Knowledge:
Before beginning this activity, students should be familiar with the following concepts:

- Regioselective addition to alkenes (Markovnikov's rule).
- Curved arrows.
- Cationic intermediates (classification as primary etc., stability).

Learning Objectives
Content Learning Objectives:
After completing this activity students should be able to:
- Draw the product and mechanism for addition of HX to an alkyne.
- Predict the products of hydrogenation and hydration of an alkyne.
- Convert between ketone and enol tautomers.

Process Objectives:
- Critical Thinking. Students analyze and evaluate different addition reactions to an alkyne, comparing the information to the addition to an alkene, to predict the products.

Class Activity 12

Electrophilic Addition to Alkynes

Model 1: Electrophilic Addition of HX to an Alkyne

Questions:

1. (a) What product is formed after addition of one equivalent of HBr (Step 1)?

   (b) In the product formed after step 1, the Br is bonded to one of the carbons of the double bond (*circle one*) true / false.

   (c) Based on your answer to #1b, complete the following sentence. The name <u>vinyl</u> bromide means the Br is ....

   (d) What product is formed after addition of the second equivalent of HBr (Step 2)?

   (e) In the product formed after Step 2, the two Br groups are attached to the (*circle one*) same / different carbon.

   (f) Based on your answer to #1e, the term germinal dibromide means the Br groups are:

2. (a) How many π-bonds does an alkyne have?

   (b) In the example in Model 1, how many π bonds have reacted in step 1? _____
      How many π bonds have reacted in step 2? _____
      How many π bonds have reacted in total? _____

   (c) Is the alkyne acting as a nucleophile or an electrophile? Explain.

   (d) Compare the reaction of the alkyne with HBr (Model 1) to the reaction of an alkene with HBr (Class Activity 11A). In your groups discuss what is similar between these reactions.

3. Consider the addition of HBr to an alkyne as shown in Model 1:
   (a) Draw the <u>two possible</u> vinyl cations (using curved arrows to show formation) after the addition of $H\oplus$ in step 1.

   (b) Of the two vinyl cations drawn above, which would be the most favorable? Explain why.

   (c) Check that the most favorable vinyl cation from #3b, leads to the vinyl bromide shown in Model 1. Does this follow Markovnikov's rule? (*Circle one*) yes / no. Explain.

   (d) Starting with the vinyl bromide, draw (using curved arrows) the <u>two possible</u> cations formed after the addition of $H\oplus$ in step 2.

   (e) Given the structure of the geminal dibromide in Model 1, discuss which of the two cations from answer #3d must be the more favorable. Explain why (HINT: use resonance forms to help explain your answer).

4. In the addition of HBr to 2-pentyne:
   (a) Draw the two possible cations that could form after the addition of $H\oplus$ in the first step.

(b)　Of the two cations drawn in #4a above, would you expect any difference in stability? Explain.

(c)　Draw the two geminal dibromides expected on reaction of 2-pentyne with excess HBr. Once you have reached agreement on the above questions, discuss whether these products will be formed in equal amounts or not.

5.　Next consider the addition of halogens to alkynes.
　　(a)　Draw the products expected on addition of 1 equivalent of Br_2 to 2-pentyne. If both SYN and ANTI addition can occur, would you expect one major product to form? Explain.

　　(b)　Draw the product(s) expected on addition of 2 equivalents of Br_2 to 2-pentyne.

Model 2: Hydrogenation of Alkynes (Addition of H_2)

Questions:

6. Consider the reaction shown in Model 2:
    (a) What functional group is formed in Eq. 1 above? _____

    (b) What functional group is formed in Eq. 2 above? _____

    (c) Describe what is different about the reagents that results in the formation of different products in Eq. 1 and Eq. 2.

    (d) Explain why Lindlar's catalyst is referred to as a "poisoned catalyst".

7. In the hydrogenation of an alkyne using Lindlar's catalyst:
    (a) If SYN addition occurs, the resulting alkene geometry would be (*circle one*) cis / trans.

    (b) Does your answer to #7a agree with the geometry shown in Eq. 2 above? (*Circle one*) yes / no.

    (c) Can the alkene in Eq. 2 be rotated so that the opposite geometric isomer is formed? (*Circle one*) yes / no. Explain.

    (d) If alkenes cannot freely rotate, discuss in your groups how the mechanism of hydrogenation would need to differ (in general terms) so that the <u>opposite</u> geometric isomer could be formed.

Model 3: Hydration of Alkynes (Addition of H_2O)

Hydration can be accomplished using either a mercury catalyst (H_2O, $Hg(OAc)_2$, H_2SO_4) or via hydroboration (R_2BH; H_2O_2, NaOH).

$$H_3C-\!\!\!\equiv\!\!\!-H \;+\; H_2O \longrightarrow \text{vinyl alcohol (enol)} \xrightleftharpoons{\text{tautomerize}} \text{ketone}$$

Questions:

8. In the reaction shown in Model 3:
    (a) The addition of water to an alkyne occurs to give a vinyl alcohol. What is another name for a vinyl alcohol? _____ Explain how this other name is derived.

(b)   The vinyl alcohol rapidly tautomerizes to form a ketone. Based on the arrows depicting this step, which compound would you expect to be the most stable?

(c)   Describe what happens in general terms in the tautomerization step.

(d)   In the hydration of propyne shown above, does the addition reaction occur according to Markovnikov's rule? (*Circle one*) yes / no.

Is this the same as hydration of alkenes? (*Circle one*) yes / no.

9. Keto-enol tautomerization is often called a proton jump as the proton on the OH group "jumps" to the neighboring carbon, as shown in Model 3.
    (a)   What bonds are being broken in the vinyl alcohol when undergoing tautomerization?

    (b)   What bonds are being formed in tautomerization to give the ketone?

    (c)   Draw the keto form of each of the following enols.

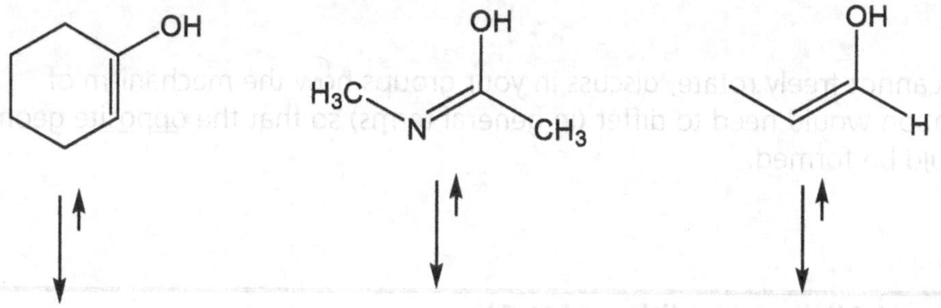

10. Below is the acid catalyzed hydration of 1-butyne.

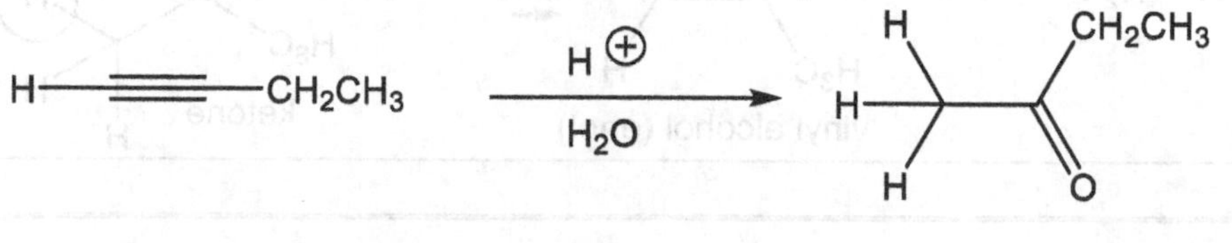

(a)   The product shown above is the (*circle one*) Markovnikov / anti-Markovnikov product.

(b)   Using your knowledge of the acid catalyzed hydration of an alkene, devise a mechanism for the above reaction, using curved arrows to show electron movement.

(c)   Draw the enol that would lead to the product drawn below. Is this enol a result of Markovnikov or anti-Markovnikov addition?

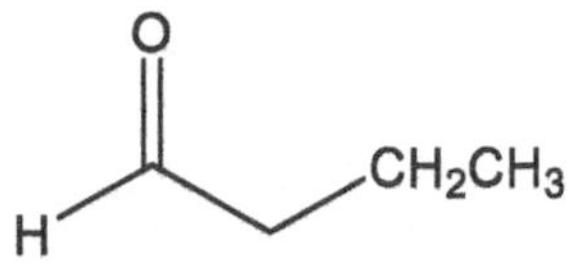

(d)   Once your group has reached agreement on the above questions, explain why the product shown in #10a is the major product and not the one shown in #10c.

(e)   From the reagents listed on the top of Model 3, determine which reagent is necessary to form the regioisomer given in #10c above.

Reflection: on a separate sheet of paper.
As a group, describe three concepts your group has learned from this activity and the one most important unanswered question about this activity that remains with your group. Turn this in before leaving class.

Additional Questions:
11. Provide reagents that would accomplish each of the following transformations.

(a)

(b)

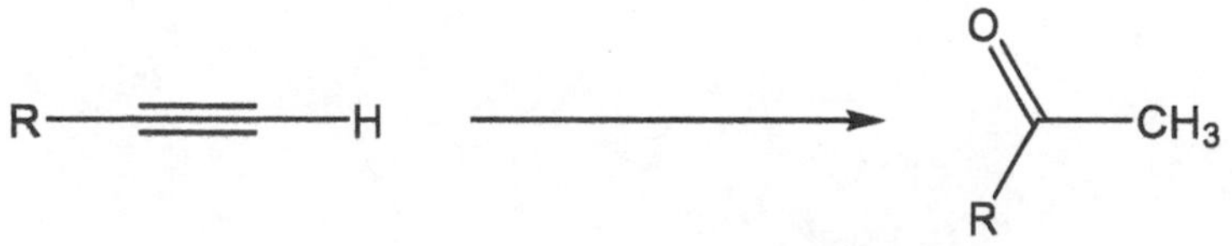

(c)

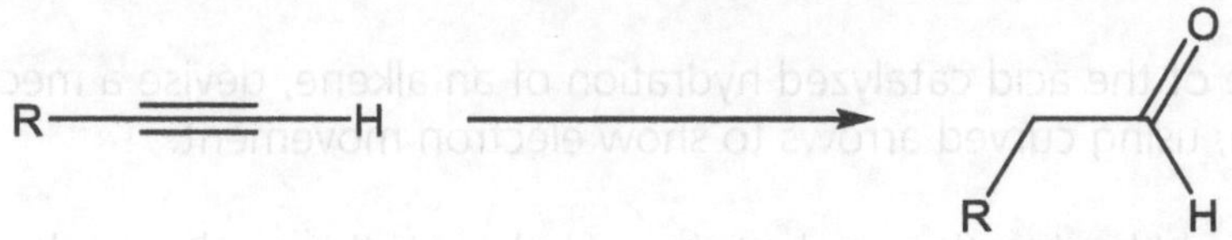

(d)

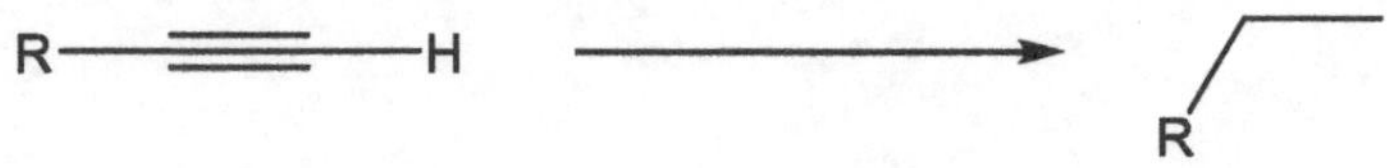

(e)

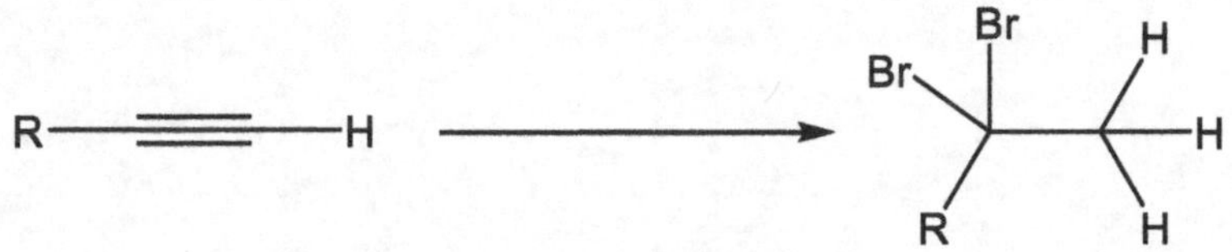

12. Indicate what reagents should be used to synthesize progesterone from the alkyne on the left.

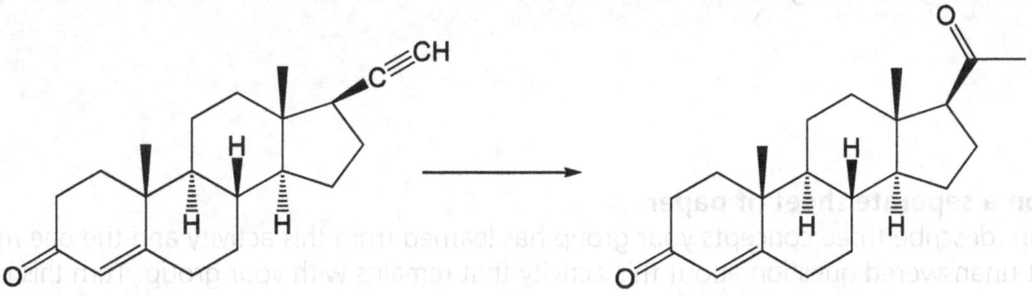

Progesterone

Class Activity 13

Organic Synthesis

Prior Knowledge:

Before beginning this activity, students should be familiar with the following concepts:

- Reactions for addition, elimination, and substitution.
- Markovnikov's rule, regiochemistry of addition.
- Curved arrows.
- Functional groups and naming.

Learning Objectives

Content Learning Objectives:

After completing this activity students should be able to:

- Propose a method for determining a synthetic pathway if only the products are known (retrosynthesis).
- Determine starting reagents from given products.
- Design two to three step syntheses of simple target molecules.

Process Objectives:

- Critical Thinking. Students explore and analyze the method of retrosynthetic analysis, integrating knowledge of reactions to determine a pathway towards target molecules.
- Problem Solving. Students plan and execute a strategy for the design of a synthesis that leads to a final product.

Class Activity 13

Organic Synthesis

Model 1: General Synthetic Strategy
Substances can be prepared (synthesized) by combining items (starting materials and reagents) in one or more steps until the target (product) is obtained.

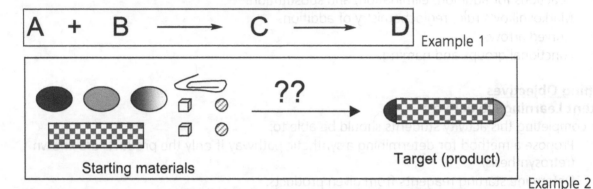

Example 1

Example 2

Questions:

1. Consider Example 1 in Model 1:

    (a) What are the starting materials?

    (b) What is the final product?

    (c) How many steps does it take to go from the starting materials to the products?

2. An alternative synthesis to that shown in Example 1 might be:

    E + F ⟶ G ⟶ D

    (a) Is the product the same as in Example 1? (*Circle one*) yes / no.

    (b) Are the starting materials the same as in Example 1? (*Circle one*) yes / no.

    (c) If there are two pathways to obtain the same target, discuss strategies that might be used to determine which pathway to follow.

3. Consider Example 2 in Model 1. The goal is to prepare an <u>exact replica</u> of the toy rattle shown (target) using a variety of supplied materials (starting materials). The amount of supplied materials is more than is needed to prepare the replica of the rattle.
    (a) You shake the rattle, and hear that something is inside. How do you know which materials to put inside the checkered tube?

    (b) Look at the egg shaped materials on the ends. Can you determine which eggs to use? Remember this must be an exact replica.

    (c) What is the best way to determine <u>exactly</u> what materials to use to prepare the replica of the rattle?

4. If you are given a target molecule with no instructions on what starting materials to use and how many steps there are, generate some ideas about how to start (what reagents and starting materials to use).

Model 2: Synthesis of an Alcohol
This same strategy as the rattle preparation can be used when designing an organic synthesis.

Questions:
5. Consider the synthesis shown in Model 2:
    (a) What is the functional group of the starting material? _____

         What is the functional group of precursor A? _____

         What is the functional group of the product? _____

    (b) How many total steps are involved in this synthesis? _____

(c)   If you did not know what the structure of precursor A is, discuss if you would be able to determine how many steps the reaction would take.

6. Consider the forward direction of the reaction (from starting material to target).
    (a)   What type of reaction occurs in Step 1? _____ What reagents might accomplish forward step 1?

    (b)   What type of reaction occurs in Step 2? _____ What reagents might accomplish forward step 2?

7. Next consider the <u>reverse</u> direction of the reaction (from target to starting material).
    (a)   Put an X through the bonds you would break if you were pulling apart a model of the TARGET to go <u>back</u> to the PRECURSOR A.
    (b)   List <u>all</u> reagents (not just those listed in #6b) that cause hydration of an alkene to form an alcohol, and note any regiochemistry associated with each reaction.

    (c)   Do any of the reagents from your list in question #7b match the reagent written above the arrow of Step 2 (from #6b)?

    (d)   Are there any reagents from your list in #7b that <u>cannot</u> be used to form the target from the precursor A? Explain why.

    (e)   How are the arrows used for the forward synthesis (making the alcohol), different from the arrows used to show the retrosynthesis or backward steps?

8. The process of working backwards from the target molecule to the starting materials is called **retrosynthetic analysis**. Thinking backwards for retrosynthesis requires knowing how different compounds can be made. In this exercise we will work backwards one step only.

   (a) Draw the starting material and reagents used to prepare each of the following alcohols. There may be more than one reagent that will accomplish the same transformation.

Product

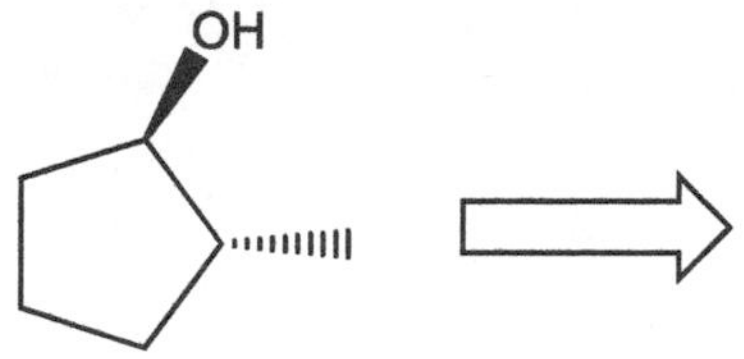

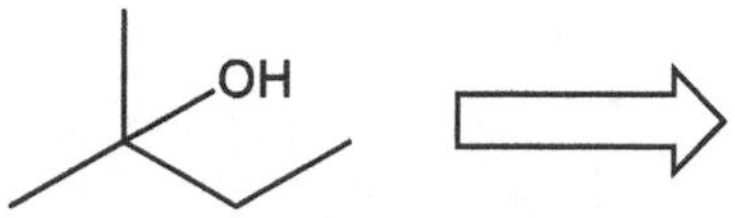

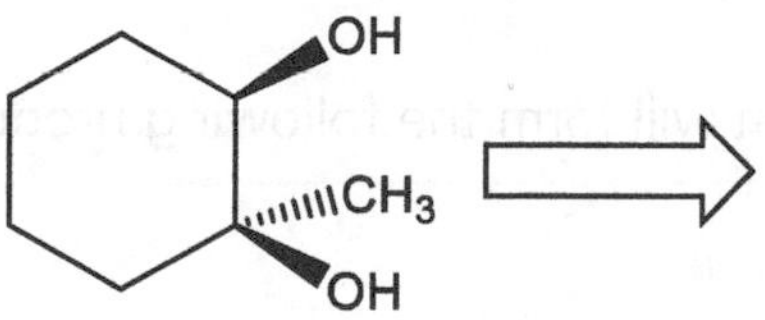

   (b) Draw the starting material and reagents used to prepare each of the following alkenes. There may be more than one reagent that will accomplish the same transformation.

Product

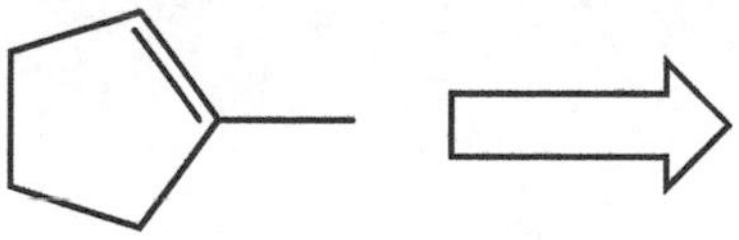

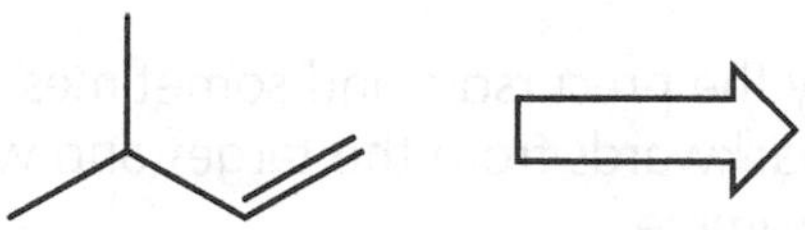

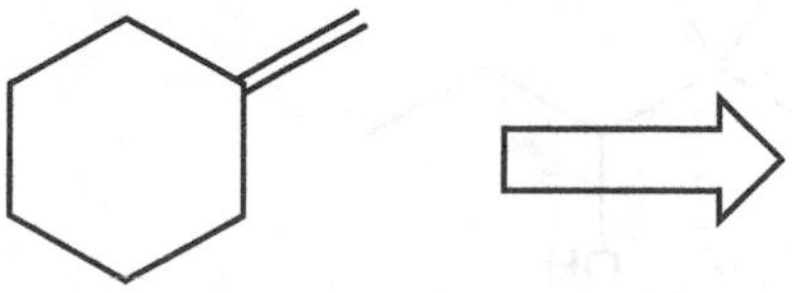

(c)   Draw the <u>alkene</u> starting material and oxidizing agent that will form the following
      oxidative cleavage products.

    <u>Products</u>

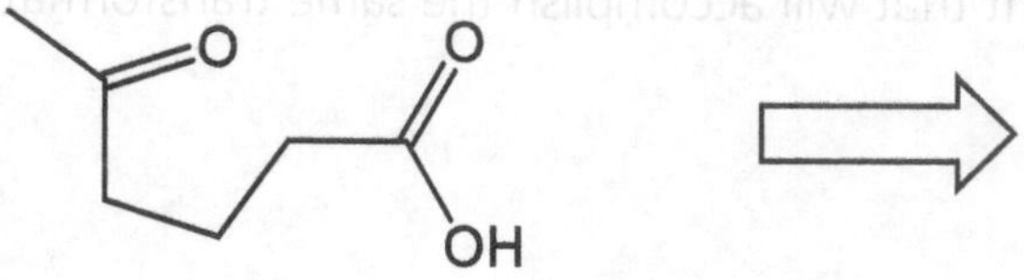

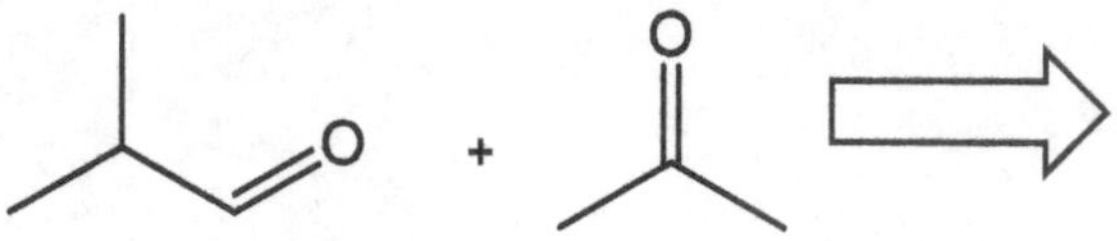

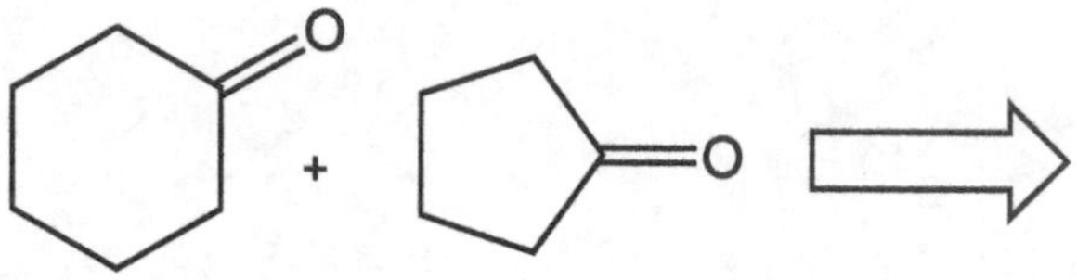

(d)   Draw the <u>alkyne</u> starting material and reagents that will form the following products.

    <u>Products</u>

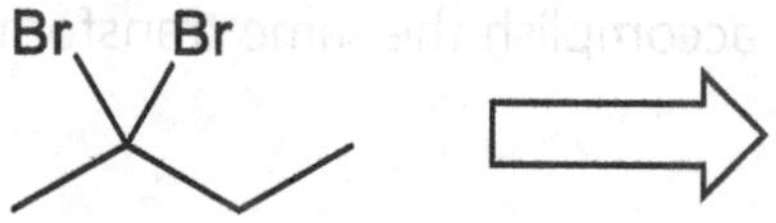

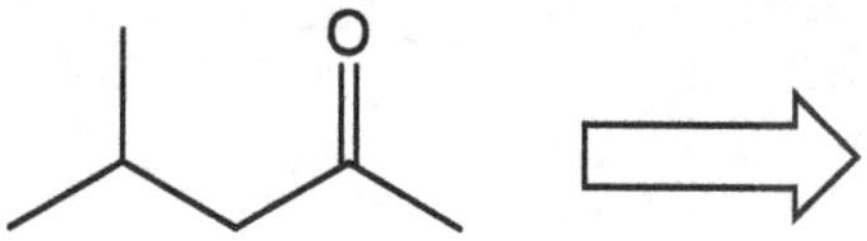

9.   Most syntheses require more than one step and usually the precursors and sometimes even the
     starting materials are not given. Thus you must think backwards from the target and work
     towards a suitable starting material. In the following example:

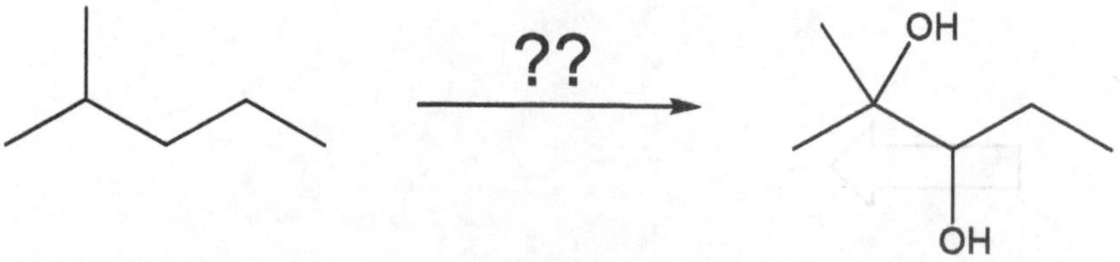

(a)   Indicate which compound is the starting material and which is the target, or product
       (HINT: look at the type of arrow drawn).

(b)   Begin with the target. Using retro arrows, draw the compound and reagents that could
       be used to form this target (it is not shown in the scheme above). This is your precursor.

(c)   Using the precursor molecule from question #9b, draw the compound and reagents that
       could be used to form this precursor, using retro arrows. Repeat this process until you get
       to the starting material shown above.

(d)   Show the entire forward synthesis, complete with reagents and precursors, that will lead
       from the starting material to the product shown above.

Reflection: on a separate sheet of paper.
    As a group, describe three concepts your group has learned from this activity and the one most
    important unanswered question about this activity that remains with your group. Turn this in
    before leaving class.

Additional Questions:

10. For the following example we will work both in the forward direction and in the backwards direction to get to a reasonable synthetic pathway.

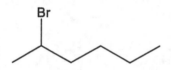

(a) Look at the starting material, and count how many carbons are present. _____

Look at the target and count how many carbons are present. _____

How many carbons must be added (in the form of a reagent) to the starting material to get the correct number of carbons in the product? _____

(b) Draw the two different reagents that the target alkyl halide could be prepared from. Is one of these routes preferable to the other one to accomplish the above synthesis? Explain.

(c) What reagent MUST be added to starting material ethyne in order to gain the correct number of carbons present in the target molecule?

(d) Draw the forward reaction of ethyne with the reagent drawn in #10c to get the precursor. Is this precursor the same compound as the one drawn in #10b?

(e)   Can the compound drawn in #10b be prepared from the precursor drawn in #10d? If so list the reagents that could be used.

(f)   Using the pathway outlined above, draw the entire synthetic sequence complete with reagents and precursors that shows formation of the target from the starting materials.

11. In the following example, using the guidelines illustrated in the above questions:

(a)   Devise a retrosynthetic plan that shows all steps leading from target to starting material.

(b)   Show the entire forward synthesis, complete with reagents and precursors, that will lead from the starting material to the product.

Class Activity 14

Organometallic Reagents

Prior Knowledge:
Before beginning this activity, students should be familiar with the following concepts:

- Electronegativity and bond polarity.
- Formal charge.
- Curved arrows.
- Nucleophile, electrophile, acid and base definitions and pKa values.

Learning Objectives
Content Learning Objectives:
After completing this activity students should be able to:
- Identify that the carbon of an organometallic reagent will be electron rich and act as the nucleophile/base.
- Predict the product when organometallic reagents are reacted with water or alcohols.
- Predict the product and draw the mechanism when organometallic reagents are reacted with carbonyl compounds.

Process Objectives:
- Information Processing. Students explore electronegativity values and how it relates to organometallic reagents and reactions.
- Problem Solving. Students identify the role of organometallic reagents and determine the products that will result when reacted with a variety of carbonyl compounds.

Class Activity 14

Organometallic Reagents

Model 1: Preparation of Grignard and Organolithium Reagents

$$R{-}\underset{\underset{R}{|}}{\overset{\overset{R}{|}}{C}}{-}Li \quad \xleftarrow{\;2\,Li\;} \quad R{-}\underset{\underset{R}{|}}{\overset{\overset{R}{|}}{C}}{-}Br \quad \xrightarrow{\;Mg\;} \quad R{-}\underset{\underset{R}{|}}{\overset{\overset{R}{|}}{C}}{-}MgBr$$

$$+\ LiBr$$

$$R = \text{alkyl or H}$$

$$\boxed{\begin{array}{c} Mg^{2+} \\ Br^{\ominus} \end{array} = MgBr^{\oplus}}$$

Questions:

1. For the reactions shown in Model 1:

   (a) What metal reacts with the alkyl halide to form the organolithium reagent?

   (b) What is the charge of a lithium ion?

   (c) What metal is reacted with the alkyl halide to form the Grignard reagent?

   (d) What is the charge on a Mg ion?

   (e) What is the charge on a Br ion?

   (f) What is the total charge on a MgBr ion?

2. (a) Consider the bond between carbon and the metal. Based on the electronegativity values (C= 2.5, Mg = 1.3, Li = 1.0), put δ+ and δ– signs on the carbon and the metal in both organometallic reagents in Model 1 to indicate the polarization.

   The carbon of the C-M bond in the organometallic reagent is (circle one) partially positive / partially negative. Explain.

   (b) Add curved arrows to the organometallic reagents in Model 1 to show how the electrons could move to break the bond between the carbon and the metal. Draw the resulting ionic compounds above.

3. Would the carbon of an organometallic reagent act as a nucleophile/base, or electrophile/acid? Explain.

Model 2: Grignard Reagent Reacting with an Alcohol

$$R\!-\!MgBr \quad + \quad \underset{CH_3}{\overset{H}{\underset{|}{\overset{|}{O}}}} \quad \longrightarrow \quad R\!-\!H \quad + \quad CH_3O\!-\!MgBr$$

Questions:

4. (a)   Draw curved arrows to show the electron movement for the reaction in Model 2.

    (b)   What is the role of RMgBr? (*Circle one*) Nucleophile / Base / Electrophile / Acid.

    (c)   What is the role of CH_3OH? (*Circle one*) Nucleophile / Base / Electrophile / Acid.

    (d)   The pKa of carbanions is around 50. Estimate the pKa value of an alcohol and predict whether the reaction in Model 2 will proceed in the forward or the reverse reaction. Explain.

    (e)   Draw the product expected on reaction of a Grignard reagent (RMgBr) with water.

    (f)   Organometallic reagents are usually prepared from alkyl halides and used directly without isolating the reactive reagent. Discuss in your groups why it is important that alcohols and water NOT be present when preparing and using organometallic reagents?

Model 3: Addition of Organometallic Reagents to Aldehydes or Ketones

$$H_3C\!:^{\ominus}\!Li^{\oplus} \;+\; \underset{H \quad\quad R}{\overset{O}{\overset{\|}{C}}} \;\;\rightleftharpoons\;\; \underset{H \quad\quad R}{\overset{H_3C \quad \ddot{\underset{\cdot\cdot}{O}}:^{\ominus}\,Li^{\oplus}}{C}} \;\;\xrightarrow{HCl}\;\; \underset{H \quad\quad R}{\overset{H_3C \quad OH}{C}} \;+\; LiCl$$

Questions:

5.   Consider the reaction shown in Model 3:

    (a)   Draw δ+ and δ– on the aldehyde to show the polarity of the C=O bond.

    (b)   Indicate which reagent is the nucleophile and which is the electrophile.

    (c)   What new bond is formed in the first step?

(d)   After acidification in the second step, what is the functional group of the final product?

(e)   Draw curved arrows to show electron movement for the reaction in Model 3.

6. (a)   Draw the product expected for the reaction of each of the following compounds with the organometallic reagent.

(b)   Identify the type of alcohol (primary, secondary or tertiary) that is formed for the three reaction in #6a.

(c)   Ketones will always give _____ alcohols. Explain.

(d)   Based on steric effects alone, discuss which would be more susceptible to reaction with a nucleophile, an aldehyde or a ketone? Explain.

(e)   Based on electronic effects (remember that alkyl groups donate electrons), discuss which would be more susceptible to reaction with a nucleophile, an aldehyde or a ketone? Explain.

Model 4: Addition of Organometallic Reagents to Acid Halides or Esters

Questions:

7. For the reaction shown in Model 4:

    (a) In the first step label the reagents as nucleophile and electrophile. What is the product formed after the first step?

    (b) Is the product formed after the first step isolated? (*Circle one*) yes / no.

    (c) In the second step label the reagents as nucleophile and electrophile.

    (d) What new bond is formed in the second step?

    (e) How many equivalents of organolithium reagents are used for the whole reaction?

    (f) What is the final product formed in Model 4?

8. (a) It has been found that the first step in the mechanism of addition to an acid halide or ester involves a tetrahedral intermediate (carbon has four single bonds). Draw the tetrahedral intermediate expected after reaction of CH₃Li with an acid halide (i.e. X=Cl).

    (b) After formation of the tetrahedral intermediate, the X group leaves to form the ketone. Draw curved arrows to show electron movement from the tetrahedral intermediate to the ketone.

    (c) After the addition of one equivalent of organometallic reagent a ketone is formed. Even if only one equivalent of nucleophile (RMgBr) is added, no ketone is isolated. In your groups discuss a reason why this might be so.

Reflection: on a separate sheet of paper.
As a group, describe three concepts your group has learned from this activity and the one most important unanswered question about this activity that remains with your group. Turn this in before leaving class.

Additional Questions
9. Design a synthesis of the following target molecule from the starting material given, using:
   (a)   any Grignard or alkyl lithium reagent
   (b)   any alkyne

   Show all reagents and steps leading to the target molecule.

   starting material   target molecule

10. Which of the following are acceptable alkyl halides for making Grignard or alkyl lithium reagents? If they are not acceptable show the side-reaction that would occur is each were treated with Li or Mg.

11. Predict the products of the following reactions

(a)

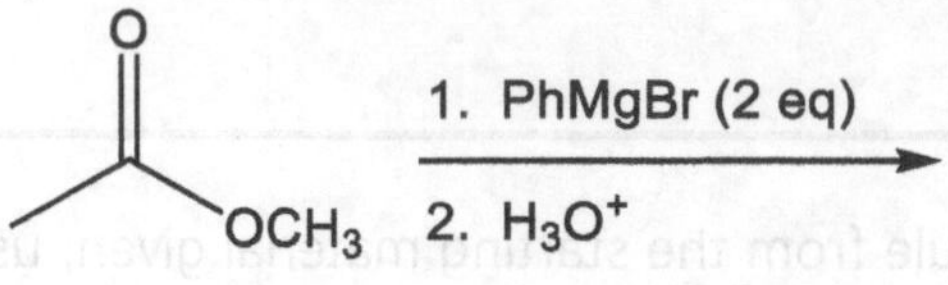

1. PhMgBr (2 eq)
2. H_3O^+

(b)

1. Mg
2.

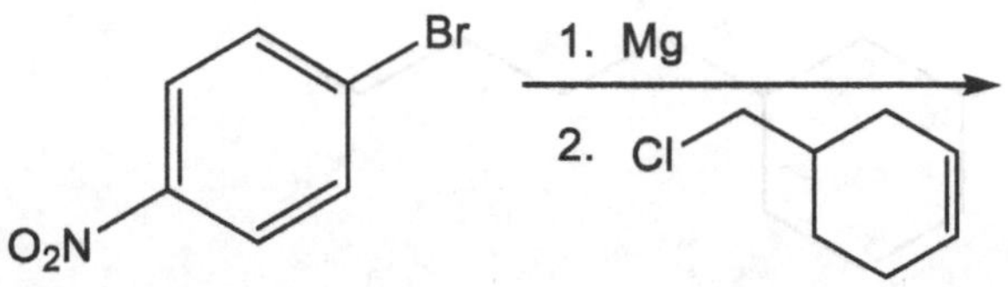

12. Grignard reactions are often used to add functionality to the "D" ring of steroids. Draw the product that would result in the steroid synthesis below.

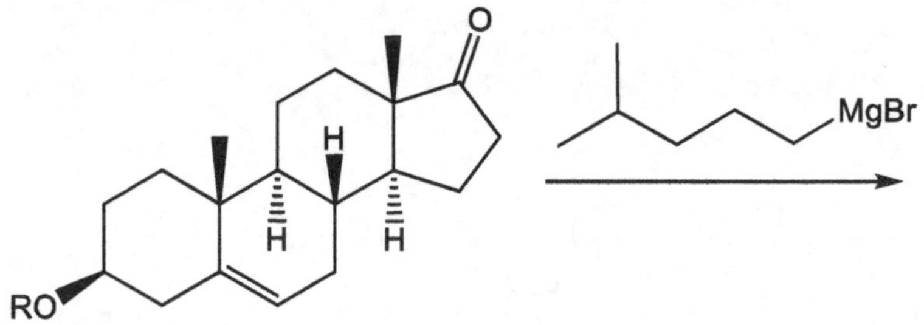

Class Activity 15

Oxidation and Reduction

Prior Knowledge:
Before beginning this activity, students should be familiar with the following concepts:

- Drawing organic structures.
- Basic organic reactions.

Learning Objectives
Content Learning Objectives:
After completing this activity students should be able to:
- Identify organic reactions as oxidation, reduction or neither.
- Predict reagents that will oxidize alcohols to various carbonyl compounds.
- Predict reagents that will reduce carbonyl compounds to various alcohols.

Process Objectives:
- Information Processing. Students explore the factors that determine whether oxidation or reduction of organic compounds has occurred.
- Critical Thinking. Students evaluate a number of oxidizing and reducing reagents to predict the products that will form.

Class Activity 15

Oxidation and Reduction

Model 1: Reduction Reactions

Reducing Reagents [Red]:
Strong Reducing Reagent: $LiAlH_4$
Weak Reducing Reagent: $NaBH_4$

$$R-CO-R \xrightarrow{[Red]} R-CH(OH)-R \quad \text{(Eq. A)}$$

$$R-CO-OH \xrightarrow{[Red]} R-CH_2-OH \quad \text{(Eq. B)}$$

$$\text{(alkene)} \xrightarrow{[Red]} \text{(alkane)} \quad \text{(Eq. C)}$$

Questions:

1. For the reduction reactions, from left to right, shown in Model 1:
    (a) The number of C-H bonds for Eq. A is (*circle one*) increased / decreased / same.
    (b) The number of C-H bonds for Eq. B is (*circle one*) increased / decreased / same.
    (c) The number of C-H bonds for Eq. C is (*circle one*) increased / decreased / same.

2. For the reduction reactions, from left to right, shown in Model 1:
    (a) The number of C-O bonds for Eq. A is (*circle one*) increased / decreased / same.
    (b) The number of C-O bonds for Eq. B is (*circle one*) increased / decreased / same.
    (c) The number of C-O bonds for Eq. C is (*circle one*) increased / decreased / same.

3. Using the above information, reduction of organic compounds can be described as an increase in the bonds to _____ or a decrease in the bonds to _____.

4. Consider the reduction reaction $CH_3CHBrCHBrCH_3 \rightarrow CH_3CH=CHCH_3$.
    (a) The number of C-H bonds is (*circle one*) increased / decreased / same.
    (b) The number of C-Br bonds is (*circle one*) increased / decreased / same.
    (c) Compare the effect of halogens to that of oxygen (#2 above) in a reduction reaction. Halogens are classified in (*circle one*) the same / differently with regards to reduction compared to oxygen.
    (d) Explain why this reaction is a reduction reaction.

5. Consider the reaction $RCHBrCH_3 \rightarrow RCH=CH_2 + HBr$

   (a) The number of C–H bonds is (*circle one*) increased / decreased / same.

   (b) The number of C–Br bonds is (*circle one*) increased / decreased / same.

   (c) Would this reaction be considered a reduction reaction? (*Circle one*) yes / no. Explain.

6. Reducing reagents can be classified as strong or weak as shown in Model 1. Compare the starting carbonyl compound to the product in each of the following reactions:

Carbonyl	LiAlH$_4$ Reductions	NaBH$_4$ Reductions
Acids	$RCO_2H + LiAlH_4 \rightarrow RCH_2OH$	$RCO_2H + NaBH_4 \rightarrow RCO_2H$
Esters	$RCO_2CH_3 + LiAlH_4 \rightarrow RCH_2OH$	$RCO_2CH_3 + NaBH_4 \rightarrow RCO_2CH_3$
Ketones	$RCOR + LiAlH_4 \rightarrow RCH(OH)R$	$RCOR + NaBH_4 \rightarrow RCH(OH)R$
Aldehydes	$RCHO + LiAlH_4 \rightarrow RCH_2OH$	$RCHO + NaBH_4 \rightarrow RCH_2OH$

   (a) Which carbonyls undergo reduction with LiAlH$_4$?
(*Circle all applicable*) acid / ester / ketone / aldehyde

   (b) Which carbonyls undergo reduction with NaBH$_4$?
(*Circle all applicable*) acid / ester / ketone / aldehyde

   (c) Are any of the reducing reagents unable to reduce some carbonyl compounds?
(*Circle one*) yes / no.

     If so, list the reagent and the carbonyl group(s) that is not reduced by the reagent.

   (d) Once everyone in your group agrees on the above concepts, determine the reducing reagents that will accomplish the following transformations:

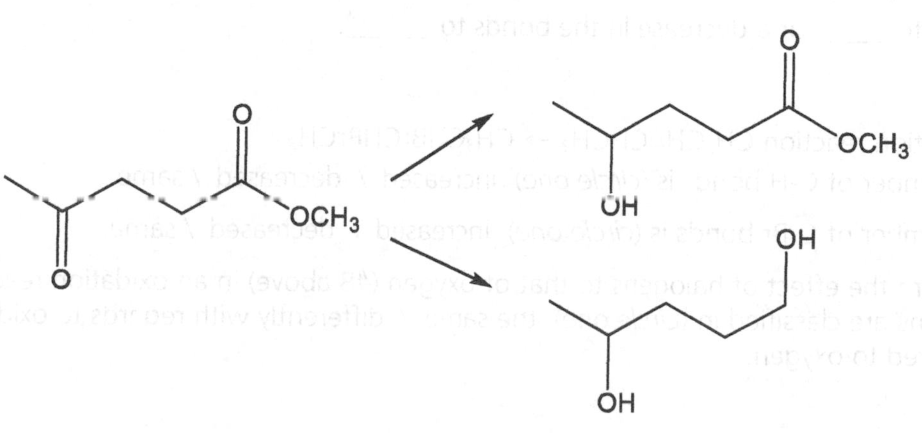

Model 2: Oxidation

$$R-CH(OH)-R \xrightarrow{[Ox]} R-C(=O)-R \quad \text{(Eq. D)}$$

$$R-CH_2-OH \xrightarrow{[Ox]} R-C(=O)-OH \quad \text{(Eq. E)}$$

$$\xrightarrow{[Ox]} \quad \text{(Eq. F)}$$

Oxidizing Reagents [Ox]:

Strong Oxidizing Reagents:
$Na_2Cr_2O_7$, H_2SO_4 or
$KMnO_4$

Weak Oxidizing Reagents:
CrO_3-Py-HCl (PCC) or
$(COCl)_2$, Et_3N, DMSO (Swern)

Questions:

7. For the oxidation reactions, from left to right, shown in Model 2:
    (a) The number of C-H bonds for Eq. D is (*circle one*) increased / decreased / same.

    (b) The number of C-H bonds for Eq. E is (*circle one*) increased / decreased / same.

    (c) The number of C-H bonds for Eq. F is (*circle one*) increased / decreased / same.

8. For the oxidation reactions, from left to right, shown in Model 2:
    (a) The number of C-O bonds for Eq. D is (*circle one*) increased / decreased / same.

    (b) The number of C-O bonds for Eq. E is (*circle one*) increased / decreased / same.

    (c) The number of C-O bonds for Eq. F is (*circle one*) increased / decreased / same.

9. Using the above information, oxidation of organic compounds can be described as an increase in the bonds to _____ or a decrease in the bonds to _____.

10. For the oxidation reaction $CH_3CH=CHCH_3 \rightarrow CH_3CHBrCHBrCH_3$
    (a) The number of C-H bonds is (*circle one*) increased / decreased / same.

    (b) The number of C-Br bonds is (*circle one*) increased / decreased / same.

    (c) Compare the effect of halogens to that of oxygen (#8 above) in an oxidation reaction. Halogens are classified in (*circle one*) the same / differently with regards to oxidation compared to oxygen.

11. Oxidizing reagents can be classified as strong or weak as shown in Model 2. Compare the starting compound to the product in each of the following reactions:

Compound	Strong [Ox] Reagents	Weak [Ox] Reagents
1° ROH	$RCH_2OH + [Ox] \rightarrow RCO_2H$	$RCH_2OH + [Ox] \rightarrow RCHO$
2° ROH	$R_2CHOH + [Ox] \rightarrow RCOR$	$R_2CHOH + [Ox] \rightarrow RCOR$
3° ROH	$R_3COH + [Ox] \rightarrow R_3COH$	$R_3COH + [Ox] \rightarrow R_3COH$
Aldehydes	$RCHO + [Ox] \rightarrow RCO_2H$	$RCHO + [Ox] \rightarrow RCHO$

(a) Strong [Ox] convert primary alcohols to _____.

(b) Weak [Ox] convert primary alcohols to _____.

(c) Strong [Ox] convert secondary alcohols to _____. Compare the oxidation of secondary alcohols with weak [Ox]. Are there any differences?
(*Circle one*) yes / no. Explain.

(d) Strong [Ox] convert aldehydes to _____. Compare the oxidation of aldehydes with weak [Ox]. Are there any differences?
(*Circle one*) yes / no. Explain.

(e) Are there any compounds not oxidized by either strong or weak [Ox]?
(*Circle one*) yes / no. If so what compounds?

(f) Once everyone in your group agrees on the above concepts, choose the reagents that will accomplish the following transformations:

Reflection: on a separate sheet of paper.
As a group, describe three concepts your group has learned from this activity and the one most important unanswered question about this activity that remains with your group. Turn this in before leaving class.

Additional Questions:

12. Classify each of the following reactions as either oxidation, reduction or neither.

(a)

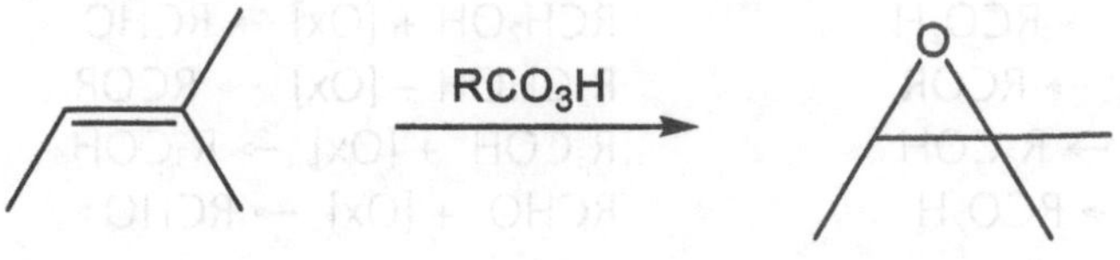
RCO3H
O

(b)

$$CH_4 + 2O_2 \longrightarrow CO_2 + 2H_2O$$

(c)

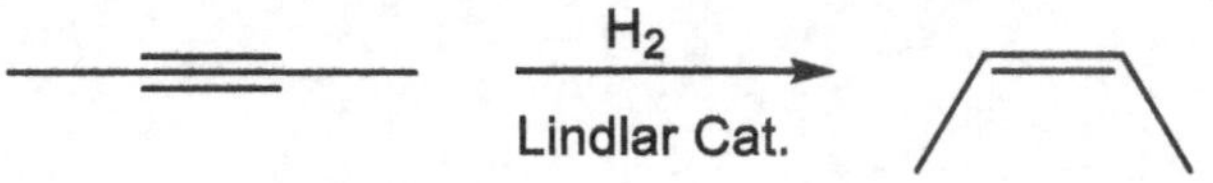
H2
Lindlar Cat.

(d)

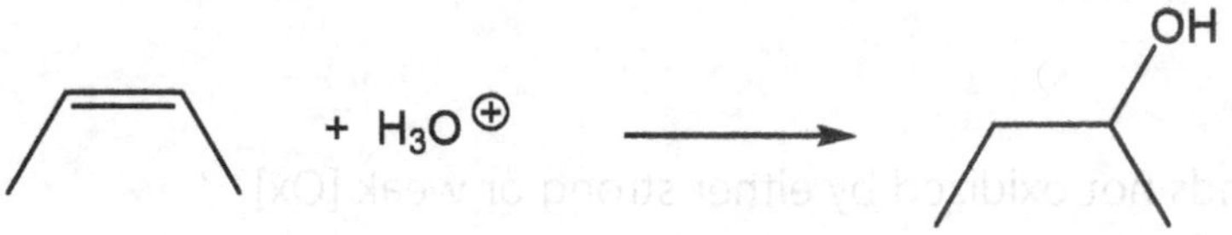
+ H3O⊕
OH

13. Shown below are some important cellular processes. Identify whether they are oxidation or reduction reactions.

(a)

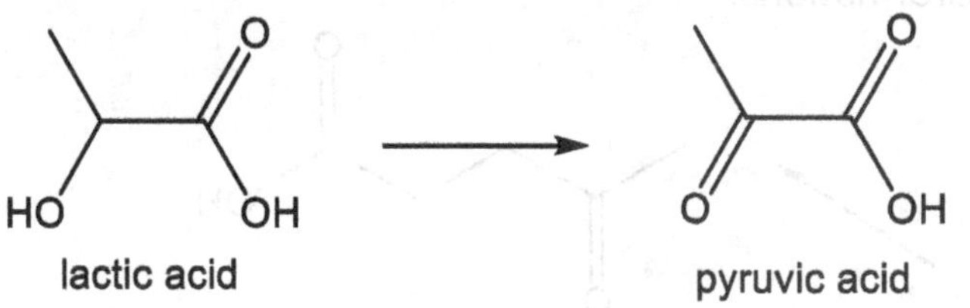
HO
OH
O
lactic acid
pyruvic acid

(b)

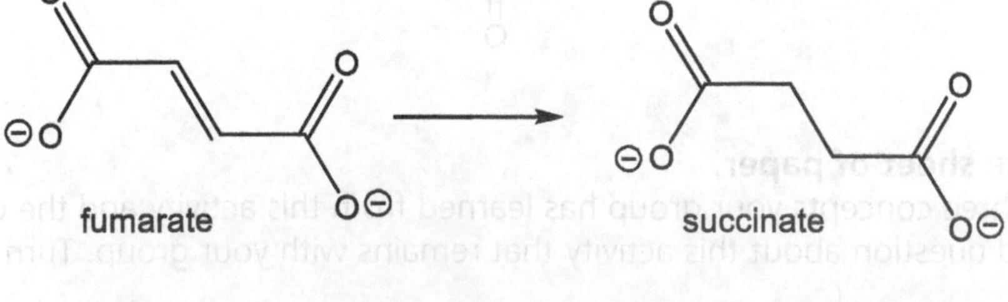
fumarate
succinate

14. Determine whether the reactions below are oxidation or reduction. List all of the reagents that will accomplish the transformations shown.

(a)

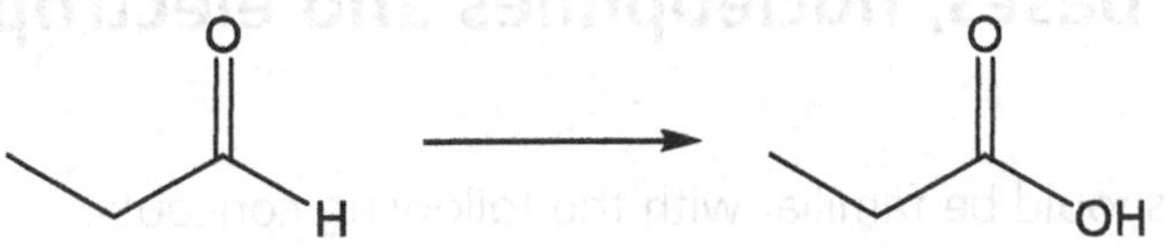

(b)

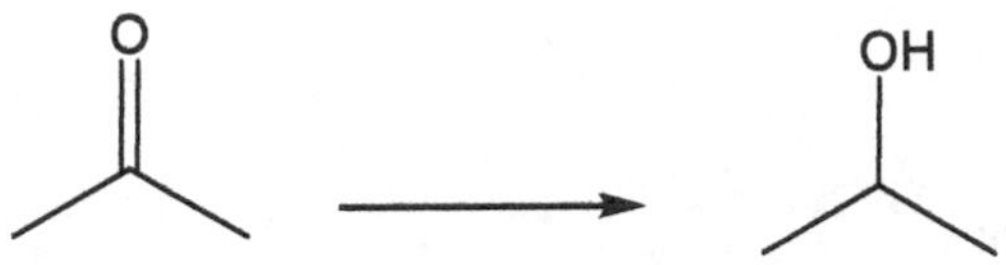

(c)

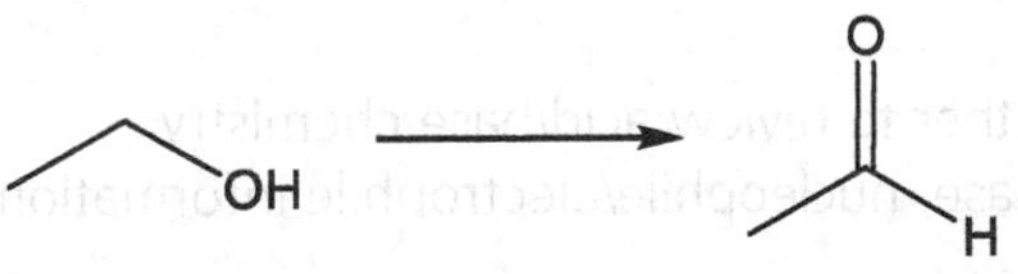

(d)

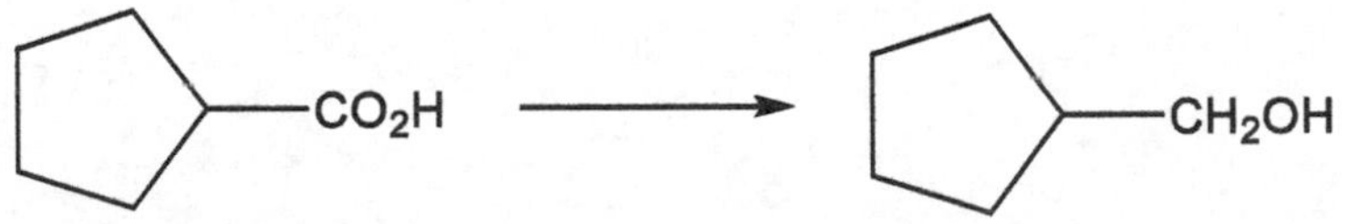

Class Activity 16A

Alcohols: Versatile Reagents
Part A: Alcohols as acids, bases, nucleophiles and electrophiles.

Prior Knowledge:
Before beginning this activity, students should be familiar with the following concepts:

- Acid/base definitions (Brönsted-Lowry, Lewis), roles and pKa values.
- Nucleophile/ electrophile definitions and roles.
- Curved arrow notation.

Learning Objectives
Content Learning Objectives:
After completing this activity students should be able to:
- Identify the acid/base/conjugate acid/conjugate base in reactions of alcohols.
- Predict whether an alcohol will act as an acid, base or a nucleophile.
- Draw curved arrows to show the mechanism of alcohol reactions.

Process Objectives:
- Teamwork. Students interact with each other to review acid/base chemistry.
- Critical Thinking. Students evaluate acid/base, nucleophile/electrophile information to draw conclusions about reactivity of alcohols.

Class Activity 16A

Alcohols: Versatile Reagents
Part A: Alcohols as acids, bases, nucleophiles and electrophiles.

Model 1: A Review of Acid / Base Reactions

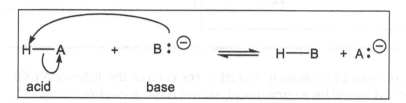

Questions:

1. Label the conjugate acid and conjugate base in the general reaction shown in Model 1.

2. Consider the acid/base reaction in terms of proton transfer:
   (a)   The acid (*circle one*) gains / loses a proton.
   (b)   The base (*circle one*) gains / loses a proton.
   (c)   The acid/base definition that refers to the gain or loss of a proton is known as the (*circle one*) Brönsted-Lowry / Lewis definition.

3. Consider the acid/base reaction in terms of electron pair transfer:
   (a)   The acid (*circle one*) donates / accepts an electron pair.
   (b)   The base (*circle one*) donates / accepts an electron pair.
   (c)   The acid/base definition that refers to an electron pair being donated or accepted is known as the (*circle one*) Brönsted-Lowry / Lewis definition.

4. (a)   Describe the differences between the two acid/base definitions in #2 and #3 above

   (b)   Look at the curved arrows drawn in Model 1. Recall that curved arrows show electron flow. Which acid/base definition helps illustrate the concept of curved arrows (*circle one*) Brönsted-Lowry / Lewis definition. Explain.

   (c)   Predict the products of the following acid base reaction. Label each reagent as acid, base, conjugate acid and conjugate base. Draw curved arrows to show electron movement.

   H——Br + HO $^{\ominus}$ $\rightleftharpoons$ H——OH + Br $^{\ominus}$

Model 2: Alcohols as Acids or Bases

Review of pKa Values:
- Strong acid pKa <5
- Strong base pKa >30
- Alcohol pKa ≈ 15

Questions:

5. Consider the review information about pKa values in Model 2. For each of the following pKa values, determine if the compound would be a strong acid, strong base or neither.

    (a)   pKa = 4   _____
    (b)   pKa = 45   _____
    (c)   pKa = -5   _____
    (d)   pKa = 20   _____

6. What is the approximate pKa of an alcohol? _____
    Based on this pKa, an alcohol would act as a (*circle one*) strong acid / strong base / neither. Explain.

7. (a)   For the reactions in Model 2, label each species as either acid, base, conjugate acid or conjugate base. Each species should have one label.
    (b)   For Eq. 1, what role does the alcohol play? (*Circle one*) acid / base.
    (c)   For Eq. 2, what role does the alcohol play? (*Circle one*) acid / base.

8. In Model 2, draw curved arrows to indicate electron movement for each reaction.

9. Draw the products of the following reaction, and identify each species as either acid, base, conjugate acid or conjugate base.

10. Once your group has reached agreement on the above questions, determine under what conditions an alcohol will act as a base and what conditions it will act as an acid. Check with a neighboring group to compare your answers.

Model 3: A Review of Nucleophiles and Electrophiles

$$E^\oplus \quad + \quad Nu\!:^\ominus \quad \rightleftharpoons \quad E\!-\!Nu$$

electrophile   nucleophile

Review Information:
➢ Nucleophiles have either a negative charge or a lone pair of electrons.
➢ Electrophiles have either a positive charge or a good leaving group.

Questions:

11. Consider the general reaction between a nucleophile and an electrophile shown in Model 3:

    (a)   A nucleophile (*circle one*) donates / accepts an electron pair.

    (b)   An electrophile (*circle one*) donates / accepts an electron pair.

    (c)   Compare the reaction in Model 3 to the reaction in Model 1. What is the same between a nucleophile and a base? What is the difference?

12. (a)   Circle compounds that are capable of acting as nucleophiles.

    $CH_3\overset{\oplus}{C}HCH_3$ $HO^\ominus$ CH_3CH_2OH CH_3Br

    (b)   Circle the compounds that are capable of acting as electrophiles.

    $CH_3\overset{\oplus}{C}HCH_3$ $HO^\ominus$ CH_3CH_2OH CH_3Br

13. The following is a substitution reaction where an alcohol replaces a halogen.

    (a)   In step 1, what bond is broken? _____ Draw curved arrows on the scheme above to show movement of the electrons in Step 1.

    (b)   What is the role of the alkyl halide (nucleophile or electrophile)? Does this agree with your definition of Nu⊖/E⊕ given in question #11? Why or why not.

    (c)   Draw curved arrows to show electron movement in Step 2.

(d)   What is the role of the alcohol in Step 2 (nucleophile or electrophile)? Does this agree with your definition of NuΘ/E⊕ given in question #11? Why or why not.

(e)   In Step 3, what bond is broken? _____ Draw curved arrows to show the electron movement for this step. This step is called deprotonation.

(f)   What is the overall mechanism of this reaction?
(*Circle one*) SN1 / SN2 / E1 / E2 / radical.

14. The net reaction shown in question #13 is outlined below:
R₃CBr + CH₃OH→ R₃COCH₃ + HBr

(a)   In general, an alkyl halide is classified as a (*circle one*) NuΘ / E⊕ / neither. Explain.

(b)   In general, an alcohol is classified as a (*circle one*) NuΘ / E⊕ / neither. Explain.

(c)   Once your group has reached agreement, come up with a plan for determining the role of the alcohol in the above reaction (whether it will act as a NuΘ or E⊕).

Reflection: on a separate sheet of paper.
As a group, describe three concepts your group has learned from this activity and the one most important unanswered question about this activity that remains with your group. Turn this in before leaving class.

Additional Questions:
15. (a)   An alcohol (pKa = 15) is reacted with NaNH₂ (pKa = 36). What role does the alcohol play (acid, base, no role)?

(b)   Draw the reaction of ROH and NaNH₂. Use curved arrows to show electron movement.

R—O—H + ⊖NH₂ ⟶
              Na⊕

(c)   Why do we define $NaNH_2$ as acting as a base and not as a nucleophile?

16. (a)   Draw the major product of the following reaction. What role does the alcohol play? (acid, base, nucleophile, electrophile). Explain.

$$CH_3OH + H_2SO_4 \rightleftharpoons$$

(b)   The reaction in #16a above is a <u>protonation</u> reaction. Consider the product $CH_3OH_2 \oplus$. Would this product react, in a subsequent step, as a nucleophile or as an electrophile? Explain.

17. (a)   Assume a nucleophile displaced a leaving group in both the alcohol and the protonated alcohol shown below. Using $Nu\ominus$ for the nucleophile, draw curved arrows to show the formation of the product and leaving group for the reaction of both alcohol and protonated alcohol.

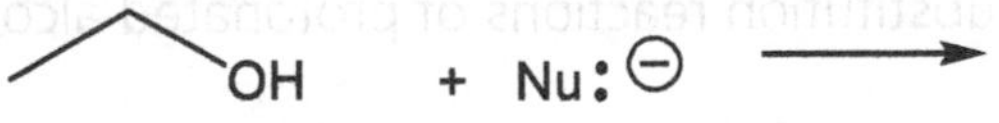

alcohol

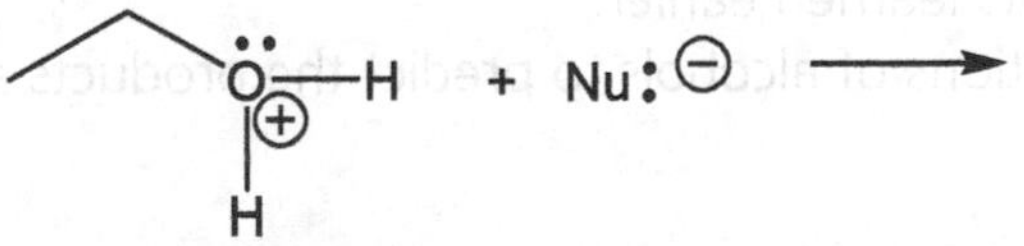

protonated alcohol

(b)   Compare the two leaving groups in the above reactions and explain which leaving group is more likely to leave and why?

18. Propranolol is used to treat hypertension (high blood pressure). Indicate which proton would be most readily abstracted under basic conditions.

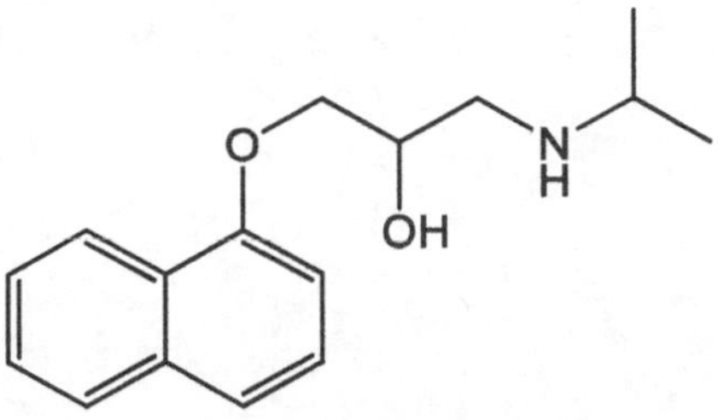

Class Activity 16B

Alcohols: Versatile Reagents
Part B: Reactions of Alcohols as Nucleophiles and Electrophiles.

Prior Knowledge:
Before beginning this activity, students should be familiar with the following concepts:

- Acid/base definitions and pKa values.
- Nucleophile/electrophile definitions and roles.
- Substitution and elimination reactions and mechanisms using curved arrows.
- Leaving group ability.

Learning Objectives
Content Learning Objectives:
After completing this activity students should be able to:
- Predict the role of an alcohol in the Williamson Ether Synthesis and Fisher Esterification.
- Determine that protonation and tosylation of an alcohol affords stronger electrophiles because of better, more stabilized, leaving groups.
- Draw the products and mechanism for substitution reactions of protonated alcohols and tosylates.

Process Objectives:
- Information Processing. Students interpret reactions of alcohols under different conditions. and relate new information with concepts learned earlier.
- Critical Thinking. Students evaluate reactions of alcohols to predict the products and mechanisms these reactions.

Class Activity 16B

Alcohols: Versatile Reagents
Part B: Reactions of Alcohols as Nucleophiles and Electrophiles.

Model 1: Reactions of Alcohols as Nucleophiles

Williamson Ether Synthesis

$$CH_3OH \xrightarrow[\text{Step 1}]{Na^o} CH_3O^{\ominus} Na^{\oplus} + \overset{}{\wedge}\!\!-Br \xrightarrow[\text{Step 2}]{} CH_3O\!\!-\!\!\wedge + NaI$$

alcohol   alkoxide   Eq. 1.

Fischer Esterification

$$H_3C\!\!-\!\!\overset{O}{\overset{\|}{C}}\!\!-\!\!OH + HO\!\!-\!\!\overset{}{\wedge} \underset{\text{heat}}{\overset{H_2SO_4}{\rightleftharpoons}} H_3C\!\!-\!\!\overset{O}{\overset{\|}{C}}\!\!-\!\!O\!\!-\!\!\overset{}{\wedge} + H_2O$$

Eq. 2

Questions:

1. Consider the Williamson Ether Synthesis, Eq. 1 of Model 1.

   (a) Circle the alcohol.

   (b) In step 1, the alcohol is converted to an alkoxide. What charge is on the oxygen of the alkoxide? _____

   (c) Which reagent (CH_3OH or $CH_3O\ominus$) would be a stronger nucleophile? Explain your answer.

   (d) In the second step, the alkoxide reacts with an alkyl halide. Label the nucleophile and the electrophile for each reagent of this second step.

   (e) The second step is a (*circle one*) substitution / elimination / addition reaction.

   (f) Draw curved arrows to show the nucleophile attacking the electrophile in Step 2.

   (g) If the alkyl halide was 2-bromo-2-methylpropane, what competing reaction would occur on reaction with the alkoxide? _____ (HINT: tertiary alkyl halide)

(h)   Draw the product and indicate what role the alkoxide plays in this reaction (nucleophile, electrophile, acid, base).

2.   Consider the Fischer Esterification, Eq. 2 of Model 1.
     (a)   Circle the alcohol. Label the carboxylic acid.

     (b)   Draw a polarity arrow on the carboxylic acid to show the polarity of the C=O bond.

     (c)   Compare the carboxylic acid (with the polarity arrow) and the alcohol. Which reagent has a partial positive charge? (*Circle one*) carboxylic acid / alcohol. Based on this information, identify what species is most likely to act as the nucleophile. Label the nucleophile and the electrophile for each reagent in Eq. 2.

     (d)   Draw curved arrows above to show the nucleophile attacking the electrophile for this reaction. (For now, don't worry about the role of the H_2SO_4 present).

     (e)   The Fischer Esterification reaction is conducted in the presence of a strong acid. Draw the competing reaction that occurs between H_2SO_4 and $NaOCH_3$. Indicate the role the alkoxide plays in this competing reaction (nucleophile, electrophile, acid, base).

     (f)   The Fischer Esterification is an equilibrium reaction. In your groups discuss how the reaction could be driven in the forward direction to form the ester in high yields.

3.   Compare the two reactions shown in Model 1. What is the role of the alcohol in each reaction? Describe the similarities and differences between these reactions.

Model 2: Conversion of Alcohols to Electrophiles

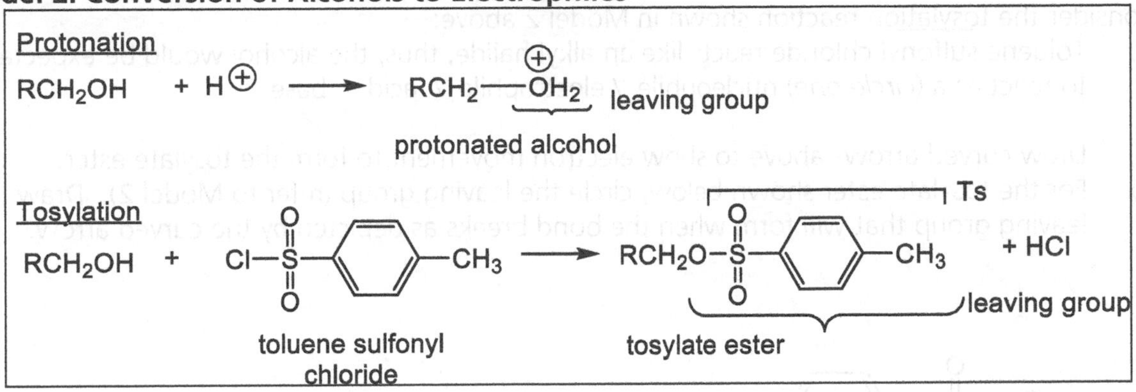

<u>Protonation</u>

RCH_2OH + $H^{\oplus}$ ⟶ RCH_2—$\overset{\oplus}{O}H_2$, leaving group

protonated alcohol

<u>Tosylation</u>

RCH_2OH + toluene sulfonyl chloride ⟶ tosylate ester + HCl

leaving group

Questions:

4. Consider the <u>protonation</u> reaction shown in Model 2 above:

   (a) What bond is being formed in the protonation reaction? _____ The alcohol is acting as a (*circle one*) nucleophile / electrophile / acid / base.

   (b) Draw curved arrows above to show electron movement to form the protonated alcohol.

   (c) For the alcohol and the protonated alcohol shown below, circle the leaving group. Draw the leaving groups that will form when the bond breaks as depicted by the curved arrow.

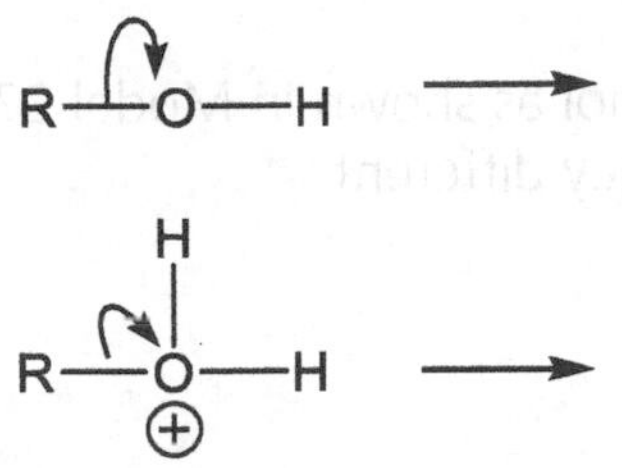

   R—O—H ⟶

   R—O(⊕)—H (with H above) ⟶

   (d) Which leaving group (as drawn in #4c) would be the best leaving group? Explain.

   (e) Based on your answers above, determine if the neutral alcohol (RCH_2OH) or the protonated alcohol ($RCH_2OH_2\oplus$) would be the stronger electrophile. Explain why.

   (f) Once your group has reached agreement on the above questions, discuss a reason for <u>why</u> acid is added to an alcohol in a reaction. Your explanation should include how the addition of acid changes the reactivity of the original alcohol.

5.   Consider the <u>tosylation</u> reaction shown in Model 2 above:
     (a)   Toluene sulfonyl chloride reacts like an alkyl halide, thus, the alcohol would be expected
           to react as a (*circle one*) nucleophile / electrophile / acid / base.

     (b)   Draw curved arrows above to show electron movement to form the tosylate ester.
     (c)   For the tosylate ester shown below, circle the leaving group (refer to Model 2). Draw the
           leaving group that will form when the bond breaks as depicted by the curved arrow.

     (d)   Draw all resonance forms of this leaving group, the tosylate anion, in the space above.

     (e)   Once your group has reached agreement on the above questions, discuss whether you
           would expect the TsO⊖ group to be a better or worse leaving group than HO⊖. Explain.

6.   What is the goal of the protonation and tosylation of an alcohol as shown in Model 2? As a
     group discuss how the reactions are the same and how are they different.

Model 3: Reactions of Alcohols as Electrophiles

$$RCH_2\overset{\oplus}{-}OH_2 + Br^{\ominus} \longrightarrow RCH_2-Br + H_2O \qquad \text{(Eq. 3)}$$

$$RCH_2-OTs + NaCN \longrightarrow RCH_2CN + NaOTs \qquad \text{(Eq. 4)}$$

Questions:
7.   (a)   Write the names <u>protonated alcohol</u> and <u>tosylate ester</u> under the corresponding reagents
           in the reactions shown in Model 3 above.

(b)   The protonated alcohol should act as a(n) (*circle one*) nucleophile / electrophile.

(c)   The tosylate ester should act as a(n) (*circle one*) nucleophile / electrophile.

(d)   Label the nucleophile and electrophile in each reaction of Model 3.

(e)   Draw curved arrows to show electron movement for both Eq. 3 and Eq. 4 above.

(f)   Draw the product of the reaction of RCH_2-Br and NaCN. Is this product the same as that obtained in Eq. 4?

(g)   Once your group has reached agreement on the above reactions, compare the reactions in Eq. 3 and Eq. 4 to the reaction of the alkyl halide in #7f. Would you expect the reactions above to follow the same mechanistic rules as alkyl halides? Explain.

Reflection: on a separate sheet of paper.

As a group, describe three concepts your group has learned from this activity and the one most important unanswered question about this activity that remains with your group. Turn this in before leaving class.

Additional Questions:

8. In each reaction below, identify the nucleophile and electrophile, and draw the product expected.

(a)

+   NaOH   ⟶

(b)

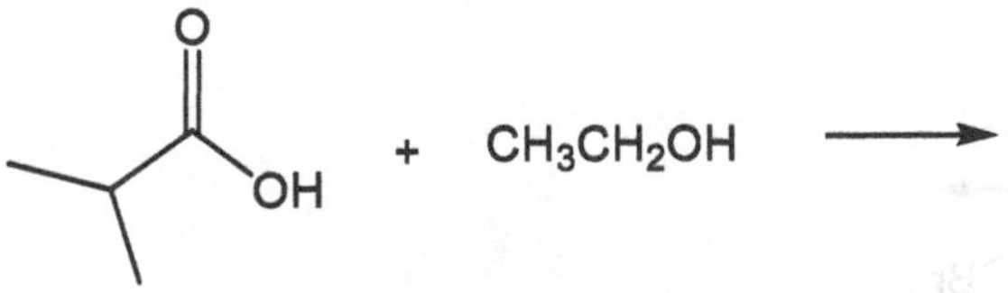

+ CH_3CH_2OH ⟶

(c)

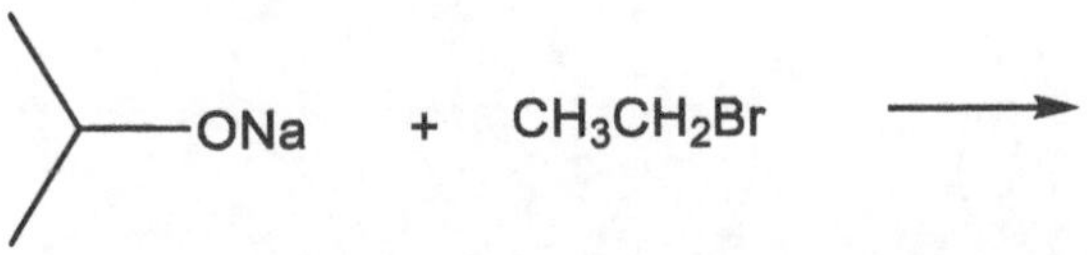

—ONa   + CH_3CH_2Br ⟶

(d)

$$CH_3CH_2OH \quad + HBr \longrightarrow$$

9. The Williamson ether synthesis is most successful when primary alkyl halides are used. Show what competing reaction can occur by drawing the product expected on reaction of NaOEt with t-butylbromide.

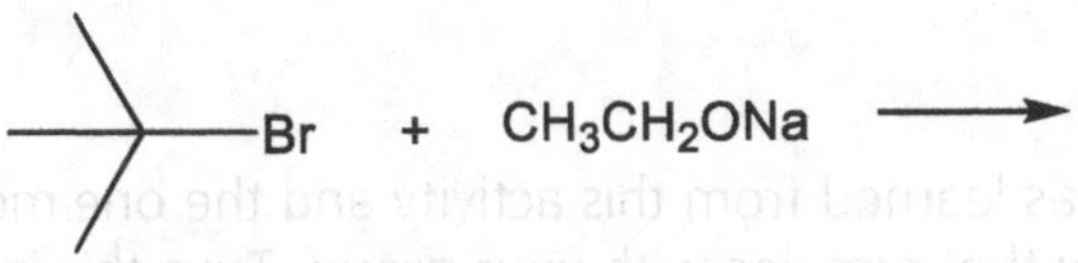

10. A reaction similar to Fischer Esterification is the reaction of an alcohol with an acid halide. Draw the products expected on the reaction of methanol with acetyl chloride ($CICOCH_3$). Identify whether the alcohol is acting as the nucleophile or the electrophile.

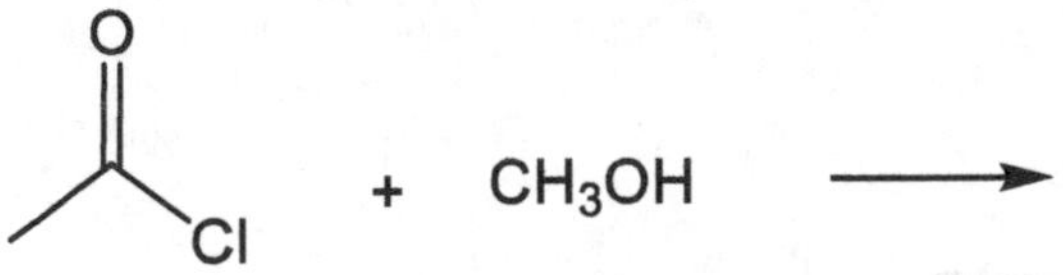

11. Acetaminophin, an analgesic, can be converted into a similar analgesic, phenacetin, using the Williamson ether synthesis as shown below. Draw the phenacetin product.

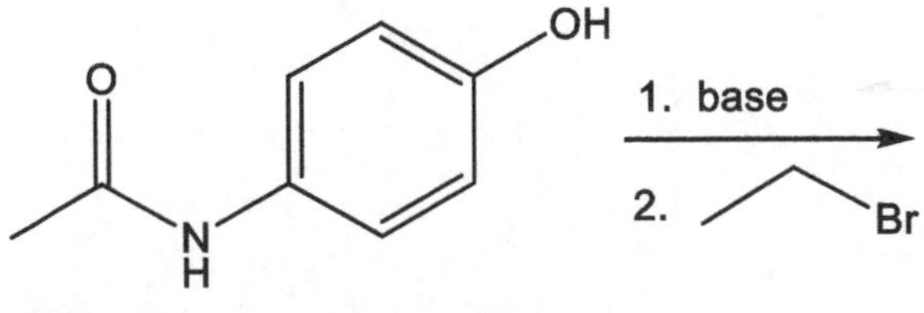

Class Activity 16C

Reactions of Diols

Prior Knowledge:

Before beginning this activity, students should be familiar with the following concepts:

- Nomenclature of alcohols.
- Protonation of alcohol, reaction and mechanism.
- Rearrangement reactions.
- Resonance stabilization.

Learning Objectives

Content Learning Objectives:

After completing this activity students should be able to:
- Establish a method for identifying diols as 1,2-diols, 1,3-diols etc.
- Draw the complete mechanism for the pinacol rearrangement.
- Predict the product of a pinacol rearrangement when the diol is unsymmetrical.

Process Objectives:

- Critical Thinking. Students analyze the reactivity of diols, using skills and concepts developed from related reactions of alcohols.
- Problem Solving. Students plan a strategy to determine the mechanism for the pinacol rearrangement.

Class Activity 16C

Reactions of Diols

Model 1: Examples of Diols

Questions:

1. How many OH groups does each compound in Model 1 contain? _____

2. Numbers are used to indicate the proximity of one OH group to another. The structure on the left in Model 1 has been numbered.

   (a) In order to determine the proximity of the OH group to each other, which carbon must be numbered 1?

   (b) For the structure on the left, if the carbon with one OH group is #1, what number is the carbon with the other OH group? _____ What general name is used to indicate the relationship between the OH groups?

   (c) Provide an IUPAC name for the 1,3-diol on the left of Model 1. How does the numbering differ in the IUPAC name compared to the general 1,3-diol name?

3. Using the compound on the left of Model 1 as an example:
   (a) Number all the remaining structures in Model 1 to determine proximity of the OH groups.

(b)   Label each compound as 1,2-diol, 1,3-diol, 1,4-diol etc.

Model 2: Pinacol Rearrangement

OH

$\xrightarrow[\text{heat}]{\text{H+}}$

+ H_2O

pinacol   pinacolone

Questions:

4. For the reaction shown in Model 2:
    (a)   Pinacol would be classified as a (*circle one*) 1,2-diol / 1,3-diol / 1,4-diol.
    (b)   Compare pinacol to the product pinacolone. What molecule has been eliminated?

    (c)   Find the longest carbon chain of pinacol, and number it as if to name the compound.
           Circle the parent chain.
    (d)   What carbon(s) contain the methyl group(s)? (give carbon #'s)

5. For the reaction shown in Model 2:
    (a)   Find the longest carbon chain of pinacolone, and number it as if to name the compound.
           Circle the parent chain.

    (b)   What carbon(s) contain the methyl group(s)? (Give carbon #'s)

    (c)   Based on your answers for #4d and #5b, discuss (in general terms) what happens to the
           methyl groups in order to convert pinacol to pinacolone.

6. Recall from Class Activity 16B that alcohols react with acid to form the protonated alcohol, as
    shown below.

OH $\xrightarrow{\text{H+}}$ $\overset{\oplus}{O}H_2$

    (a)   What is the purpose of this protonation step?

    (b)   Pinacol is reacted with acid. What role does pinacol play? (*Circle one*) acid / base.

(c)   Would you expect the first step in the reaction of pinacol with acid to be the same as the protonation of an alcohol as shown above? (*Circle one*) yes / no. Explain.

(d)   Draw the product of the first step in the reaction of pinacol with acid. What is the purpose of this step? (Refer to #6a).

OH

OH

$\xrightarrow[\text{heat}]{\text{H+}}$

pinacol

7. (a)   Once pinacol is protonated (in #6d), draw the cation that would form when the leaving group eliminates. Use curved arrows to show electron movement.

(b)   Consider the product pinacolone. What group needs to migrate in order for the cation to be converted to pinacolone? (*Circle one*) H / CH_3 / OH.

(c)   Draw a curved arrow showing migration of the above group towards the electron deficient carbon. Draw the resulting cation.

(d)   Compare the two cations drawn above, the cation initially formed and the cation formed after migration. Is either cation resonance stabilized? (*Circle one*) yes / no.

If so, draw any possible resonance structures. Can a complete octet be formed with either cation? (*Circle one*) yes / no.

(e)   Once your group agrees on the above questions, discuss if there are any differences in stability between the initial cation and the cation formed after migration. Would the cation formed after migration be more or less stable than the original? Explain.

8. (a)   Draw any resonance forms of the cation shown below.

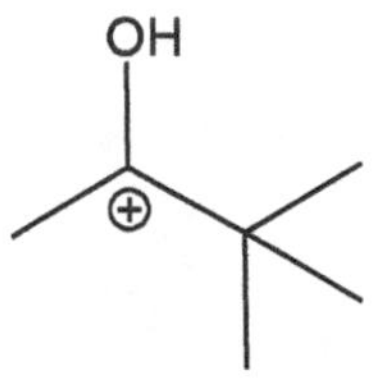

   (b)   Number the carbons in the cation above from left to right, as if to name. Does this numbering match the numbering of pinacolone? (*Circle one*) yes / no.

   (c)   What remaining step is left in order to convert this cation to pinacolone?

   (d)   Draw curved arrows to illustrate this final step to form pinacolone.

9.   Once your group has reached agreement on the above questions, determine if a different product would result if the other OH group was protonated in the first step of this mechanism. Explain your answer.

10.   Consider 3-methyl-2-phenylbuta-2,3-diol (shown below).

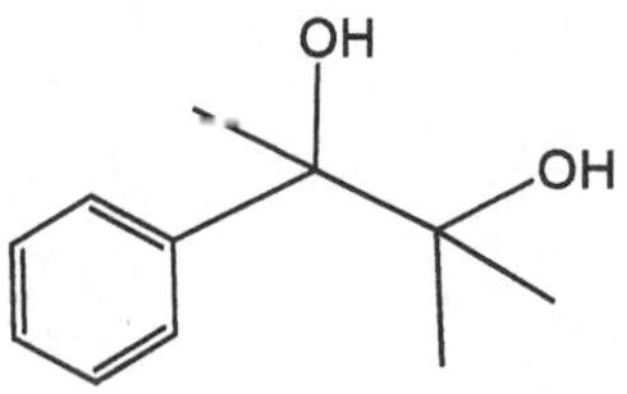

   (a)   Is this compound a 1,2-diol? (*Circle one*) yes / no.

(b)   Draw the cation formed if the OH group at C-2 is protonated.

(c)   Draw the cation formed if the OH group at C-3 is protonated.

(d)   Compare the two cations from #10b and #10c. Is there any difference in stability of these two cations? (*Circle one*) yes / no. Explain your answer.

(e)   Generally the lower energy intermediate is most likely to be formed. Based on this information draw the product formed on reaction of 3-methyl-2-phenylbuta-2,3-diol with acid.

Reflection: on a separate sheet of paper.
    As a group, describe three concepts your group has learned from this activity and the one most important unanswered question about this activity that remains with your group. Turn this in before leaving class.

Additional Questions:
11. Draw the product of the following pinacol rearrangements
(a)

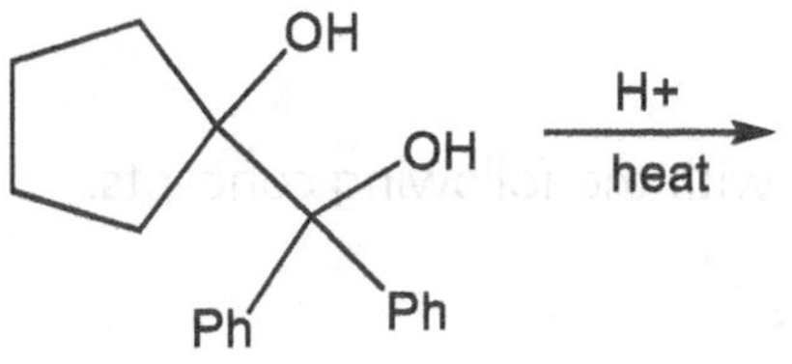

(b)

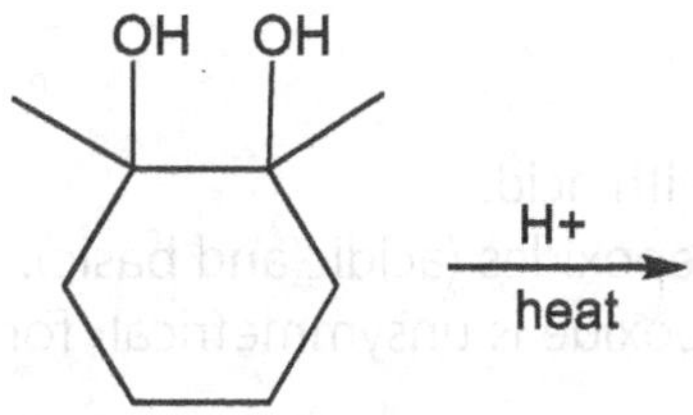

12. Draw the complete mechanism to account for the following reaction.

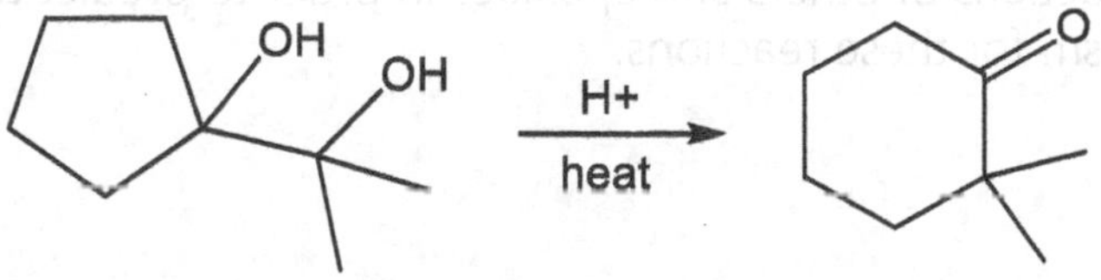

Class Activity 17

Reactions of Ethers and Epoxides

Prior Knowledge:

Before beginning this activity, students should be familiar with the following concepts:

- Acid, base, nucleophile and electrophile properties.
- Protonation of alcohol, reaction and mechanism.
- SN1, SN2, E1, E2 reactions and mechanisms.
- Stereochemistry.

Learning Objectives

Content Learning Objectives:

After completing this activity students should be able to:

- Draw the product and mechanism for ether cleavage with acid.
- Draw the product and mechanism for ring opening of epoxides (acidic and basic).
- Predict the regiochemistry of ring opening when the epoxide is unsymmetrical, for both acidic and basic conditions.

Process Objectives:

- Information Processing. Students manipulate concepts of acids, bases, nucleophiles, electrophiles, protonation and substitution reactions.
- Critical Thinking. Students analyze reactions of ethers and epoxides in order to predict the products and determine the mechanism for these reactions.

Class Activity 17

Reactions of Ethers and Epoxides

Model 1: Ether Cleavage with HBr

$$R\text{---}\overset{..}{\underset{..}{O}}\text{---}R' + H\text{---}Br \xrightarrow{\text{Step 1}} R\text{---}\overset{\overset{H}{|}}{\underset{..}{O}}\text{---}R' + Br^\ominus \xrightarrow{\text{Step 2}} ROH + R'Br$$

ether

$$\xrightarrow[\text{HBr}]{\text{Step 3}} RBr + H_2O$$

Questions:

1. For step 1 of the reaction shown in Model 1, the ether is reacted with HBr.
    (a) HBr is a (*circle one*) strong / weak acid.
    (b) In the presence of HBr, the <u>ether</u> acts as a(n) (*circle one*) acid / base.
    (c) Draw curved arrows to show electron movement for step 1.
    (d) Draw the product when an alcohol is reacted with acid. (Hint: recall activity 16B).

    $RCH_2OH + H\oplus \rightarrow$

    Compare this to the reaction of an ether with acid (step 1). Are there any differences? Explain.

    (e) Step 1 is called protonation. What is the purpose of protonation (for either an alcohol or an ether)?

2. For step 2 of the reaction shown in Model 1:
    (a) The protonated ether reacts with Br⊖. Label the roles of each reagent above (nucleophile, electrophile, acid or base).
    (b) Draw curved arrows to show electron movement for step 2.
    (c) Circle the leaving group from the reagents shown in step 2.
    (d) Assuming that the R' group is primary, by what mechanism does the reaction in step 2 occur (SN1, SN2, E1, E2, addition)? Explain.

3. In step 3 of the reaction shown in Model 1, the alcohol reacts with an additional equivalent of HBr to form an alkyl bromide. Draw the complete mechanism for step 3.

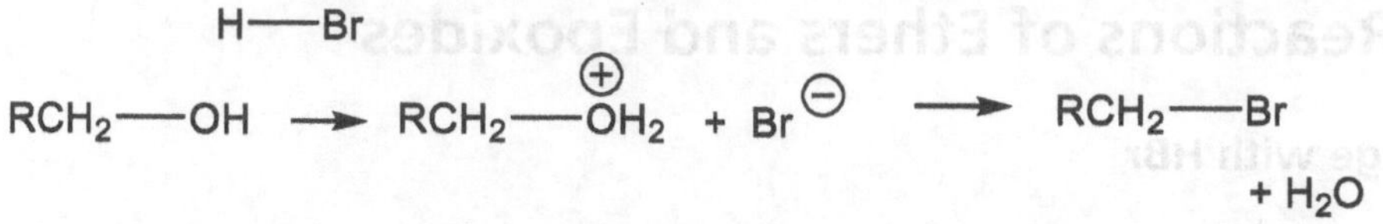

$$RCH_2{-}OH \xrightarrow{\quad H{-}Br \quad} RCH_2{-}\overset{\oplus}{O}H_2 \;+\; Br^{\ominus} \longrightarrow \begin{array}{l} RCH_2{-}Br \\ + H_2O \end{array}$$

4. The reaction of phenyl ethyl ether with HBr is shown below.

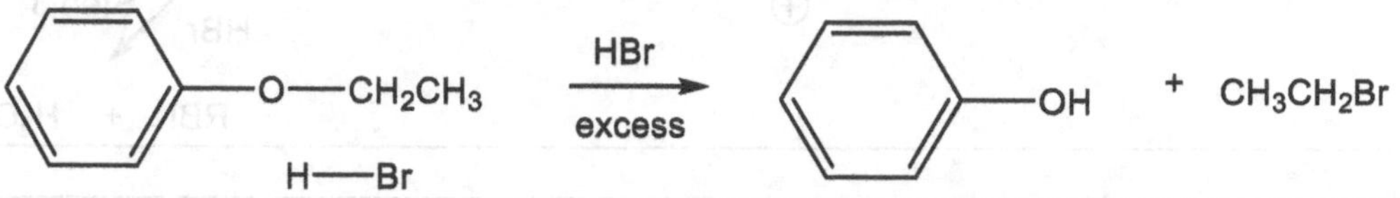

(a) What are the functional groups of the two products that are formed?

(b) Compare these products to the <u>final</u> products obtained in Model 1. Are there any differences in the products between these two reactions? If so, note the differences.

(c) Draw the mechanism above, using curved arrows to show electron movement.

(d) Once your group has reached agreement on the above questions, determine whether the phenol would react further with HBr to form PhBr. (HINT: substitution reactions do not occur at sp_2 hybridized centers). Explain.

Model 2: Ring Opening Reactions of Epoxides

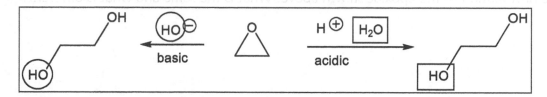

Questions:
5. (a) Label the epoxide in Model 2.
   (b) What product forms on reaction of the epoxide under acidic conditions? _____

   What product forms on reaction of the epoxide under basic conditions? _____

   Are there any differences between the products obtained under acidic vs basic conditions? _____

   (c) Compare the molecular formula of the epoxide to the formula of the products. What has been added to the epoxide?

   (d) What is the role of the water (boxed) in acidic conditions?
        (*Circle one*) nucleophile / electrophile.

   What is the role of the hydroxide (circled) in basic conditions?
        (*Circle one*) nucleophile / electrophile.

   (e) Epoxides are cyclic ethers. In your groups determine why epoxides are more reactive than ethers.

6. Consider the <u>basic</u> catalyzed opening of epoxides shown in Model 2:
   (a) What reagent is added to the epoxide?

   (b) Label the roles of each reagent involved (nucleophile vs electrophile).

   (c) Draw curved arrows to show electron movement in this step.

7. Consider the <u>acidic</u> catalyzed opening of epoxide in Model 2:
   (a) Draw curved arrows to show electron movement for <u>Step 1</u> of the mechanism shown below.

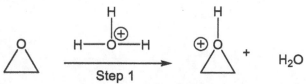

(b)   Draw the mechanism for the protonation of an alcohol (ROH + H₃O⊕), and compare to the protonation of the epoxide shown above. What is the same and what is different?

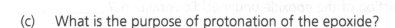

(c)   What is the purpose of protonation of the epoxide?

(d)   In the second step of the mechanism (shown below), identify the nucleophile and the electrophile. Draw curved arrows to illustrate this step.

(e)   What process must occur to convert the product of Step 2, to the final product diol?

(f)   Once your group has reached agreement on the above questions, draw the product that would form if the acid catalyzed opening of epoxides was conducted in acid and <u>methanol</u>.

Model 3: Regiochemistry of Reactions with Unsymmetrical Epoxides

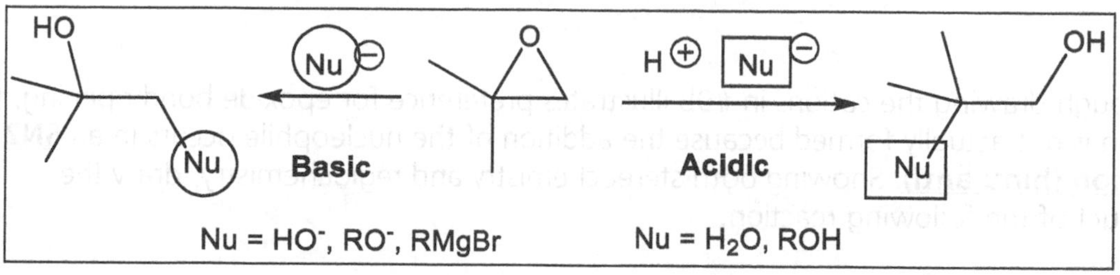

Nu = HO⁻, RO⁻, RMgBr   Nu = H₂O, ROH

Questions:

8. Refer to the reactions in Model 3:

   (a)   Under <u>acidic</u> conditions, the nucleophile attacks the (*circle one*) more / less substituted carbon of the epoxide. Under acidic conditions the nucleophiles are (*circle one*) weak / strong.

   (b)   Under basic conditions, the nucleophile attacks the (*circle one*) more / less substituted carbon of the epoxide. Under basic conditions the nucleophiles are (*circle one*) weak / strong.

   (c)   If Nu⊖ is the same for acidic and basic conditions, the relationship between the two products would be (*circle one*) constitutional isomers / conformational isomers / same / not isomers.

9. Consider the acid catalyzed opening of an unsymmetrical epoxide.

   (a)   The epoxide reacts with acid to form the protonated epoxide. The protonated epoxide would be (*circle one*) more / less reactive than the neutral epoxide. Based on your answer, explain what kind of nucleophiles (strong or weak) are necessary under acidic conditions.

   (b)   The protonated epoxide is shown below. Draw the <u>cation</u> that would be formed if 'bond a' of the epoxide was broken and the cation that would be formed if 'bond b' was broken in the ring opening reaction.

(c)   Which bond of the epoxide is more likely to be broken under acidic conditions? Explain why.

10. (a)   Although drawing the cations in #9b illustrates preference for epoxide bond opening, the cation is not actually formed because the addition of the nucleophile occurs in an **SN2 fashion (hint: anti)**. Showing both stereochemistry and regiochemistry, draw the product of the following reaction.

H ⊕

CH₃OH

(b)   After your group agrees on the product of the above reaction, consider if the epoxide above was reacted HBr. What species would act as the nucleophile?

Draw the product that would result.

11. Consider the base catalyzed opening of an unsymmetrical epoxide.
   (a)   Under basic conditions the _neutral_ epoxide reacts with a nucleophile. Explain what kind of nucleophiles (strong or weak) are necessary under basic conditions.

   (b)   Draw curved arrows to show electron movement in the reaction shown below.

CH₃O ⊖

O ⊖   H

OCH₃

H

   (c)   Given that the reaction mechanism is a concerted, SN2 mechanism, explain why the nucleophile attacks the least substituted carbon of the epoxide.

(d)   At the carbon where attack of the nucleophile occurs, what stereochemical outcome would be expected, (*circle one*) retention / inversion of configuration?

Reflection: on a separate sheet of paper.
As a group, describe three concepts your group has learned from this activity and the one most important unanswered question about this activity that remains with your group. Turn this in before leaving class.

Additional Questions:

12. Draw the major organic product of each of the following reactions, clearly showing stereochemistry if necessary. Indicate whether the reaction occurs under acidic or basic conditions. Practice drawing the complete mechanisms for each of these reactions.

(a)

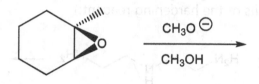

(b)

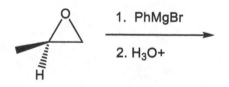

(c)

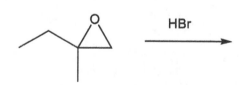

(d)

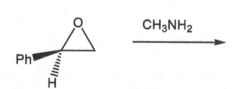

(e)

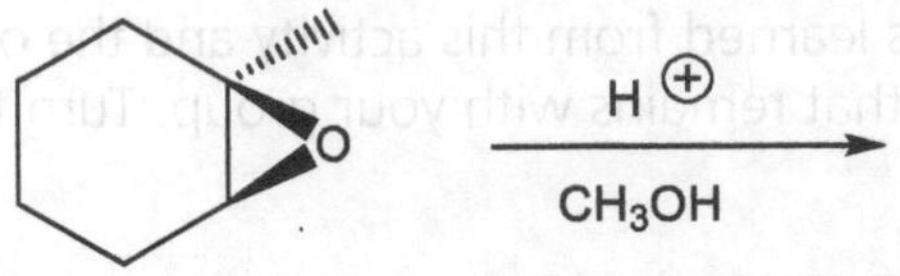

13. Epoxy resins are glues that are made from combining a prepolymer, which contains an epoxide group, with a hardening reagent, usually an amine. The prepolymer drawn below is abbreviated to only show the reactive epoxide groups at the end of the polymer. Draw the product expected when the prepolymer reacts with the hardening reagent. (HINT: your product should include two units of the prepolymer and two units of the hardening reagent.)

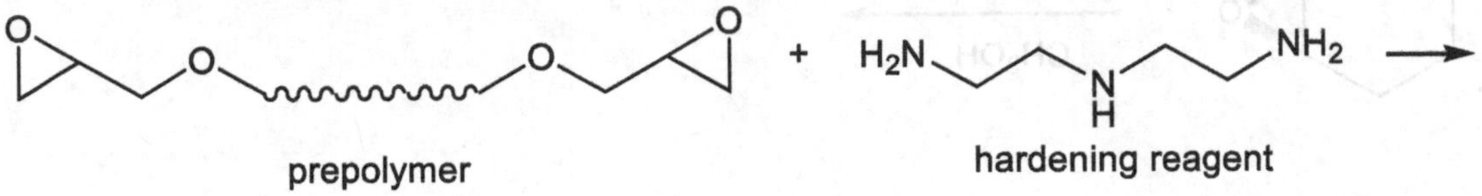

prepolymer   +   hardening reagent

Class Activity 18

Diels-Alder Reaction

Prior Knowledge:
Before beginning this activity, students should be familiar with the following concepts:

- Nucleophile and electrophile properties.
- Electron withdrawing and donating groups and effect of resonance.
- Alkene geometry and conjugation in dienes.
- Stereochemistry.

Learning Objectives
Content Learning Objectives:
After completing this activity students should be able to:
- Predict the Diels-Alder products, with correct regiochemistry and stereochemistry.
- Determine the dienes and dienophiles that will be reactive in a Diels-Alder reaction.
- Determine the Diels-Alder starting reagents, when given the product.

Process Objectives:

- Critical Thinking. Students analyze and evaluate information about polarity (electron withdrawing or donating groups), bonding and stereochemistry in order to discern how the Diels-Alder reaction proceeds.

Class Activity 18

Diels-Alder Reaction

Model 1: Diels-Alder Reaction

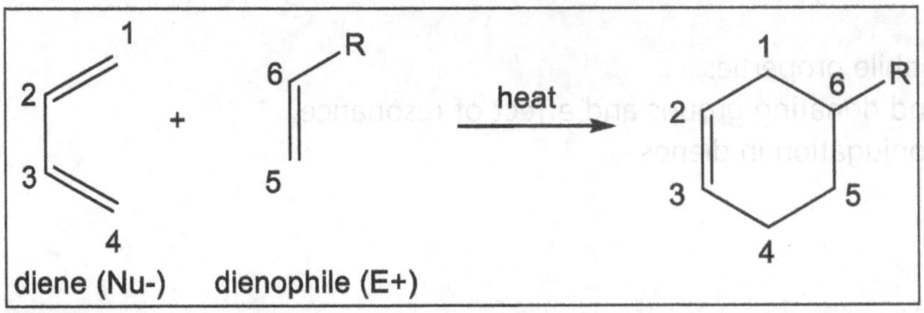

diene (Nu-)   dienophile (E+)

Questions:
1. For the Diels-Alder reaction shown in Model 1:
   (a)   The diene acts as the (*circle one*) nucleophile / electrophile.
   (b)   The dienophile acts as the (*circle one*) nucleophile / electrophile.
   (c)   For the reaction between the diene and the dienophile, indicate what bonds are broken and what bonds are formed. (Use the numbers given to indicate bond location).
         Broken: _____   Formed: _____

   (d)   Draw curved arrows to show the bonds formed and broken in the Diels-Alder reaction above. The Diels-Alder reaction occurs by a <u>concerted</u> mechanism. What does this mean?

   (e)   Using dotted lines to show partial bonds (for each bond that is broken and formed in the reaction), draw the transition state for the Diels-Alder reaction.

   (f)   How many π-electrons are involved in the Diels-Alder reaction?

2. You should be able to work backwards from the <u>product</u> of a Diels-Alder reaction and determine which diene and dienophile were used.
   (a)   Refer to the numbering of the <u>product</u> in Model 1. What numbers do the carbon atoms of the double bond have? _____

(b)   The following cyclohexene is a product of a Diels-Alder Reaction. Number the carbon atoms of the cyclohexene from 1-6, keeping in mind the numbering established in #2a.

(c)   Put an X through each single bond that was formed in the above compound, as a result of the Diels-Alder reaction (refer to #1b).

(d)   Disconnect these single bonds and draw the diene and the dienophile used to form the above product.

(e)   Once your group has reached agreement on the above questions, draw the diene and the dienophile used to form the following Diels-Alder product.

Model 2: Dienes and Dienophiles

Questions:

3. Consider the <u>dienes</u> shown in Model 2:
    (a)   What role does the diene play in the Diels-Alder reaction? (HINT: see Model 1)
          (*circle one*) nucleophile / electrophile
    (b)   Based on the role chosen in #3a, a diene should be (*circle one*) electron rich / electron deficient.
    (c)   The dienes shown are all (*circle one*) conjugated / isolated / cumulated.
    (d)   The dienes shown are all in the (*circle one*) s-cis / s-trans conformation.

4. The geometry of the diene in a Diels-Alder reaction must be in the s-cis conformation.
    (a) Which of the following dienes are capable of undergoing a Diels-Alder reaction?

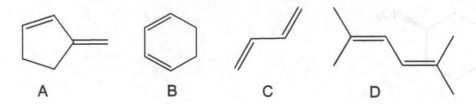

                A   B   C   D

    (b) One of the above dienes is <u>not</u> reactive in a Diels-Alder reaction. Which diene would be unreactive and why?

    (c) Compare dienes C and D. Which diene would react faster? Explain why.

5. Consider the <u>dienophiles</u> shown in Model 2:
    (a) What role does the dieneneophile play in the Diels-Alder reaction?
         (*circle one*) nucleophile / electrophile (HINT: see Model 1)
    (b) Based on the role chosed in #5a, a dienophile should be (*circle one*) electron rich / electron deficient.

6. (a) What kind of <u>substituent</u> would enhance the electrophilic nature of the dienophile?
         (*Circle one*) electron donating / electron-withdrawing. Explain.

    (b) Are all of the substitutents on the dienophiles shown in Model 2, electron withdrawing groups? (*Circle one*) yes / no.
    (c) Circle the groups that would be considered to be electron withdrawing groups.

        $-OCH_3$, $-CN$, $-CO_2R$, $-NO_2$

(d) Draw all the resonance contributors for the alkene below (at least 3 more). Once your group agrees, discuss how these resonance contributors help explain your answer to #6a.

Model 3: Stereospecific SYN Addition in Diels-Alder Reaction

Questions:

7. (a) In Model 3, label the diene and dienophile.
   (b) The geometry of the <u>dienophile</u> is (circle one) cis / trans.
   (c) The stereochemistry of the product shown in Model 3 is (circle one) cis / trans.
   (d) As a group discuss what happens to the geometry of the dienophile as it goes from starting material to product in a Diels-Alder reaction.

   (e) If a trans substituted dienophile is reacted with butadiene, what would be the orientation of the product? (Circle one) cis / trans.

8. Draw the product of the following Diels-Alder reaction, clearly showing stereochemistry.

Model 4: Stereoselective Endo Addition in the Diels-Alder Reaction

The Diels-Alder reaction is stereoselective, because reaction of <u>cyclic</u> dienes occurs to give one major stereoisomer. The endo product is favored because it allows maximum overlap of the π-electrons of the R group with the π-bonds of the diene.

Questions:

9. For the reaction shown in Model 4:
   (a)   Label the diene and dienophile.
   (b)   Number the carbons of the diene (#'s 1-4) and the carbons of the dieneophle (#'s 5-6). (use Model 1 as the reference).
   (c)   In the product, number the carbons of the six membered ring (1-6).

   (d)   What is the difference between the endo and the exo product in Model 4?

   (e)   Which product is the major product? (*Circle one*) endo / exo.
   (f)   Draw the two possible products of the following reaction. Circle the major product.

Reflection: on a separate sheet of paper.

As a group, describe three concepts your group has learned from this activity and the one most important unanswered question about this activity that remains with your group. Turn this in before leaving class.

Additional Questions:

10. Draw the products for each of the following Diels-Alder reactions

(a)

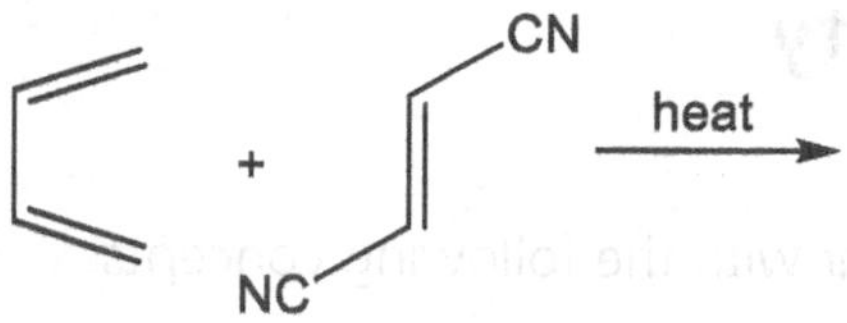

(b)

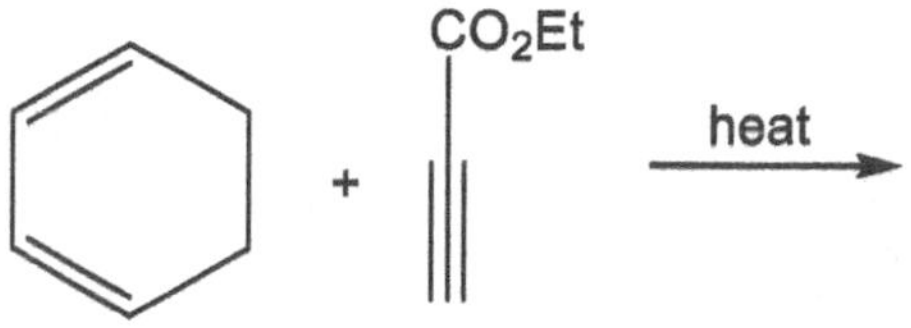

(c)

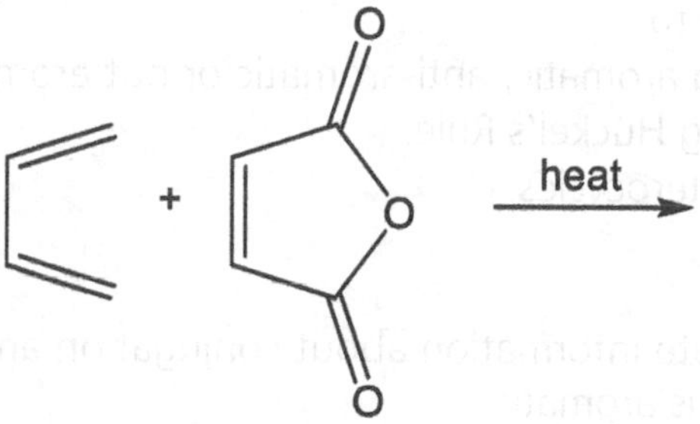

11. The Diels-Alder reaction is commonly used in the synthesis of the six membered rings contained in steroids such as cortisone, shown below.

(a) Which double bond of the dienophile would be more reactive in the Diels-Alder reaction? Explain.

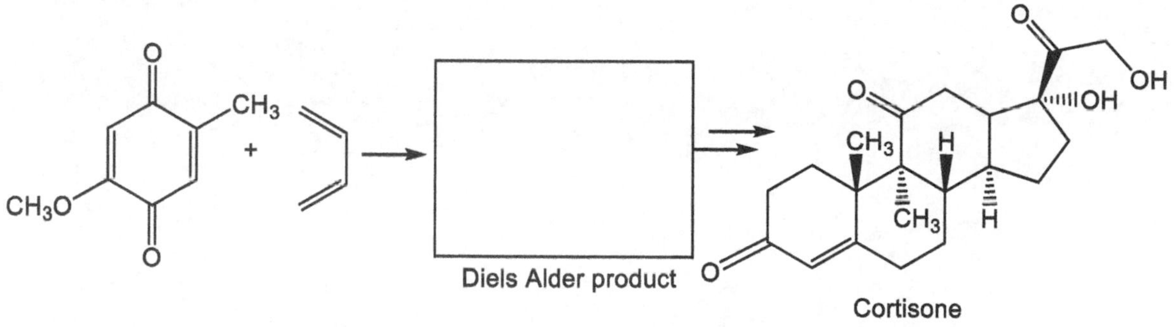

(b) Draw the product of the initial Diels-Alder reaction, clearly showing stereochemistry.

Class Activity 19

Aromaticity

Prior Knowledge:
Before beginning this activity, students should be familiar with the following concepts:

- Electrons, bonding and orbitals.
- Conjugation.
- Heat of hydrogenation.

Learning Objectives
Content Learning Objectives:
After completing this activity students should be able to:
- Determine the factors that make a compound aromatic, anti-aromatic or not aromatic.
- Predict whether a compound is aromatic using Hückel's Rule.
- Apply the rules for aromaticity to ions and heterocycles.

Process Objectives:
- Critical Thinking. Students analyze and evaluate information about conjugation and pi electrons to determine whether a compound is aromatic.

Class Activity 19

Aromaticity

Model 1: Aromatic and Non-Aromatic Compounds

Aromatic Compounds

Non-Aromatic Compounds

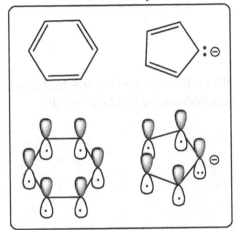
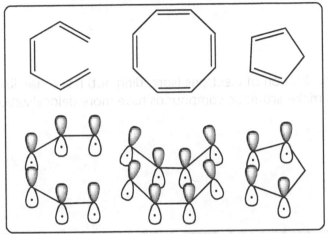

NOTE: Each "•" is one electron in a p-orbital.

Questions:

1. For the <u>aromatic</u> compounds shown on the left in Model 1.
   - (a)   Is every compound cyclic? (*Circle one*) yes / no.
   - (b)   Is every compound conjugated? (*Circle one*) yes / no.
   - (c)   Is every atom with a p-orbital connected to atoms on both sides that have a p-orbital? (*Circle one*) yes / no.
   - (d)   Based on the orbital drawings, are all p-orbitals in the same plane? (*Circle one*) yes / no.

2. For the <u>non-aromatic</u> compounds shown on the right in Model 1:
   - (a)   Is every compound cyclic? (*Circle one*) yes / no.
   - (b)   Is every compound conjugated? (*Circle one*) yes / no.
   - (c)   Is every atom with a p-orbital connected to atoms on both sides that have a p-orbital? (this is called continuously conjugated) (*Circle one*) yes / no.
   - (d)   Based on the orbital drawings, are all p-orbitals in the same plane? (*Circle one*) yes / no

3. Once everyone in your group agrees on the above questions, list the features that make a compound aromatic.

4. (a) Addition of 1 mole of H_2 to cyclohexene releases -28.6 kcal. Estimate the energy that should be released to add 3 moles of H_2 to cyclohexatriene (benzene).

    (b) Addition of 3 moles of H_2 to cyclohexatriene (benzene) actually releases -49.8 kcal. What conclusions can you make about the stability of benzene?

5. Delocalization of electrons (spreading out) has a stabilizing effect. Discuss and list the features that make aromatic compounds have more delocalization than non-aromatic compounds.

Model 2: Hückel's Rule

Hückel's rule defines the number of π electrons involved in aromatic systems. Aromatic systems have 4n+2 π electrons, while anti-aromatic systems have 4n π electrons, where n is an integer.

n	Aromatic (4n+2 π electrons)	Anti-Aromatic (4n π electrons)
0	2	0
1	6	4
2	10	8
3	14	12
4		

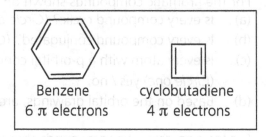

Benzene   cyclobutadiene
6 π electrons   4 π electrons

Questions:

6. (a) How many π electrons does benzene have? _____

    Based on Hückel's rule, benzene would be (*circle one*) aromatic / anti-aromatic.

    (b) How many π electrons does cyclobutadiene have? _____

    Based on Hückel's rule, cyclobutadiene would be (*circle one*) aromatic / anti-aromatic.

    (c) Complete the table for when n=4.

7. Ions can also be aromatic if they are cyclic, continuously conjugated, planar, and the number of π electrons obeys Hückel's Rule. For each of the following compounds, determine the number of π electrons, classify it as aromatic, anti-aromatic or not aromatic, and give a reason for your conclusion.

Compound	# π electrons	Classification	Reason

Model 3: Aromatic Heterocycles

Heterocycles, compounds containing O, N and S, can also be aromatic. Lone pair electrons on these atoms must be part of the delocalized π system in order to be counted for Hückel's Rule.

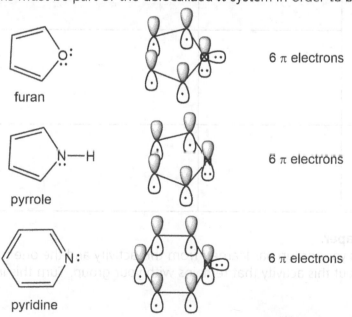

furan   6 π electrons

pyrrole   6 π electrons

pyridine   6 π electrons

Questions:

8. For the compounds shown in Model 3:

   (a)   How many π electrons does furan have?

   (b)   Given that there are two lone pairs on oxygen, is one pair or both pairs of electrons counted as being part of the delocalized system? Explain why that might be, using the orbital pictures.

   (c)   How many π electrons does pyrrole have? ___ Is the lone pair of electrons on N counted as being part of the delocalized system for pyrrole?

   (d)   How many π electrons does pyridine have? ____ Is the lone pair of electrons on N counted as being part of the delocalized system for pyridine?

   (e)   Explain why there is a difference in counting the lone pair of electrons on nitrogen for pyrrole and pyridine, using the orbital pictures shown.

9. For each of the following compounds, determine the number of π electrons, classify it as aromatic, anti-aromatic or not aromatic, and give a reason for your conclusion.

Compound	# π electrons	Classification	Reason

Reflection: on a separate sheet of paper.

As a group, describe three concepts your group has learned from this activity and the one most important unanswered question about this activity that remains with your group. Turn this in before leaving class.

Additional Questions:

10. Recall that aromatic compounds must be cyclic, continuously conjugated, planar and have 4n+2 π electrons. The following compound is not aromatic. Determine which of the above requirements are not met and discuss why.

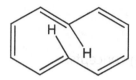

11. Determine if the following compounds are not aromatic, aromatic or anti-aromatic.

12. The following compound readily eliminates CO_2 to form a conjugated six membered ring.

$$-CO_2 \longrightarrow$$

    (a)   Draw the structure of this compound and explain why the reaction occurs so readily.

    (b)   The starting material can be formed via a Diels-Alder reaction. Draw the diene and dieneophile that could be used to prepare this starting material.

Class Activity 20A

Electrophilic Aromatic Substitution (EAS)
Part A: Reaction of Benzene

Prior Knowledge:
Before beginning this activity, students should be familiar with the following concepts:

- Energy diagrams, endothermic, exothermic, activation energy.
- Conjugated and isolated alkenes.
- Cationic intermediates, stability and formation.
- Addition to alkenes.
- Acid, base, nucleophile and electrophile roles and properties.
- Curved arrow notation.
- Resonance structures.

Learning Objectives
Content Learning Objectives:
After completing this activity students should be able to:
- Articulate why benzene is less reactive than cyclohexene.
- Draw the mechanism for electrophilic aromatic substitution.
- Predict the products of electrophilic aromatic substitution.

Process Objectives:
- Information Processing. Students interpret combined information from the prior knowledge.
- Critical Thinking. Students analyze aromatic substitution reactions and draw conclusions about the mechanism.

Class Activity 20A

Electrophilic Aromatic Substitution (EAS)
Part A: Reaction of Benzene

Model 1: Addition of Br₂ to Cyclohexene and Benzene
The energy diagram for two addition reactions (Rxn. 1 and Rxn 2) is shown below.

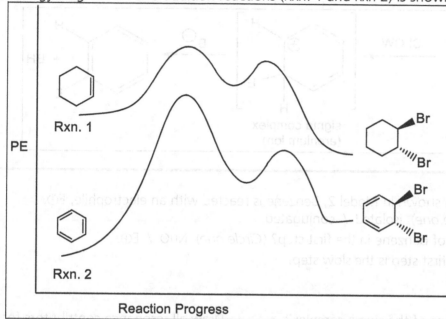

Questions:

1. For Reaction 1 in Model 1:
   (a)   The starting alkene is (*circle one*) isolated / conjugated.
   (b)   The energy diagram for Rxn 1 shows that the reaction is (*circle one*) endothermic / exothermic.

2. For Rxn 2 in Model 1:
   (a)   The starting alkene is (*circle one*) isolated / conjugated.
   (b)   The energy diagram for Rxn 2 shows that the reaction is (*circle one*) endothermic / exothermic.

3. (a)   Which starting alkene is lower in energy? (*Circle one*) benzene / cyclohexene.
   (b)   Which reaction has the higher activation energy? (*Circle one*) Rxn 1 / Rxn 2.

(c) Rxn 2 does not occur. In your groups, discuss why the reaction of benzene (Rxn 2) is unlikely compared to Rxn 1.

Model 2: Electrophilic Aromatic Substitution (EAS)

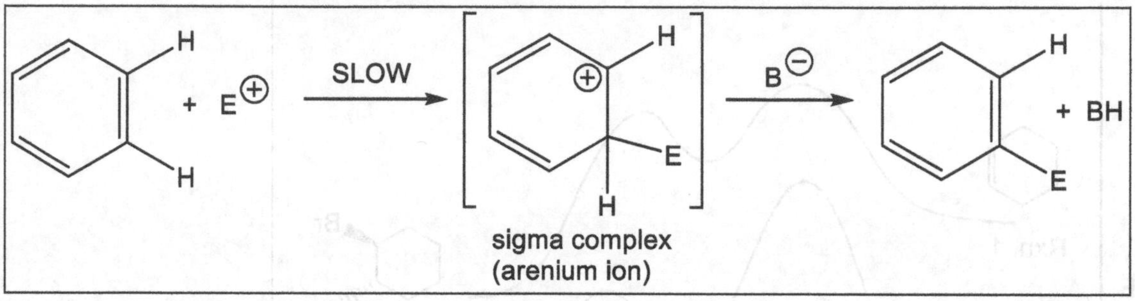

Questions:

4. In the general reaction shown in Model 2, benzene is reacted with an electrophile, E⊕.
   (a) Benzene is (*circle one*) isolated / conjugated.
   (b) What is the role of benzene in the first step? (*Circle one*) Nu⊖ / E⊕.
   (c) Explain why the first step is the slow step.

5. (a) What is the charge of the sigma complex? ______ Draw all resonance contributors for the sigma complex.

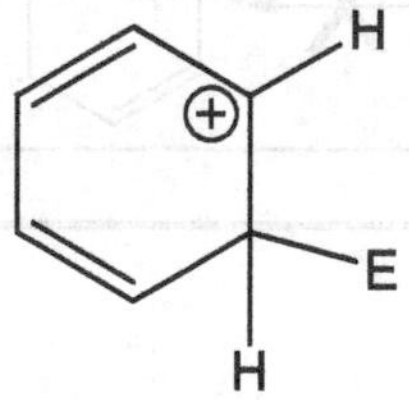

sigma complex
(arenium ion)

   (b) In the second step the base abstracts a proton to form the substitution product. What is the role of the sigma complex? (*Circle one*) Nu⊖ / E⊕ / acid / base.

   (c) As a group discuss whether this second step would occur readily with a <u>weak</u> base. Explain.

6. Draw curved arrows to show electron movement for the reaction in Model 2.

$$\text{benzene} + E^\oplus \xrightarrow{\text{SLOW}} \left[\text{sigma complex} \right] \xrightarrow{B^\ominus} \text{product} + BH$$

**sigma complex
(arenium ion)**

Model 3: Electrophiles in Electrophilic Aromatic Substitution Reactions

In order to react with a stable aromatic compound like benzene, the electrophile must be very reactive. The table below lists the activated electrophiles and the reagents used to form them.

Nu⊖	E⊕	Reagents	Product	Reaction Name
benzene	Br⊕ Cl⊕	Br₂, FeBr₃ Cl₂, AlCl₃	X X=Cl,Br	Halogenation
	NO₂⊕	HNO₃, H₂SO₄	NO₂	Nitration
	SO₃	SO₃, H₂SO₄	SO₃H	Sulfonation
	R⊕	RX, AlCl₃	R R = alkyl	Friedel-Crafts alkylation
	(acylium)	R—C(O)—X, AlCl₃	C(O)R	Friedel-Crafts acylation

Questions:

7. All EAS reactions occur by the same general mechanism as outlined in Model 2. The only difference lies in formation of the activated electrophiles from the reagents listed in Model 3. Consider halogenation, the first entry in the table from Model 3.

   (a) For bromination, what is the actual electrophile (E⊕)? _____

What are the <u>reagents</u> used to generate the electrophile? _____

What is the electrophile and reagents for chlorination? _____.

(b)   For halogenation to occur, a Lewis Acid must be used as one of the reagents. What is the general definition of a Lewis acid?

(c)   In the list of reagents for halogenation, what are the Lewis acids used?

8. (a)   When Br_2 reacts with $FeBr_3$, an activated bromine complex is formed (shown below). Draw curved arrows to show electron movement when forming this activated complex. Put $\oplus$ and $\ominus$ charges on the compound as appropriate.

$$Br\!-\!Br \;+\; FeBr_3 \longrightarrow \left[\, Br\!-\!Br\!-\!FeBr_3 \,\right]$$

(b)   The activated bromine complex formed in #8a is a strong electrophile and can react with benzene as outlined in Model 2. As a group determine which Br in the activated complex acts as the electrophile and accepts electrons.

(c)   Draw curved arrows to show electron movement in the first step of the EAS mechanism. Put a $\oplus$ charge in the correct position on the sigma complex, and draw any additional resonance forms. Do your arrows show attack of the electrophile determined in #8b above? Explain.

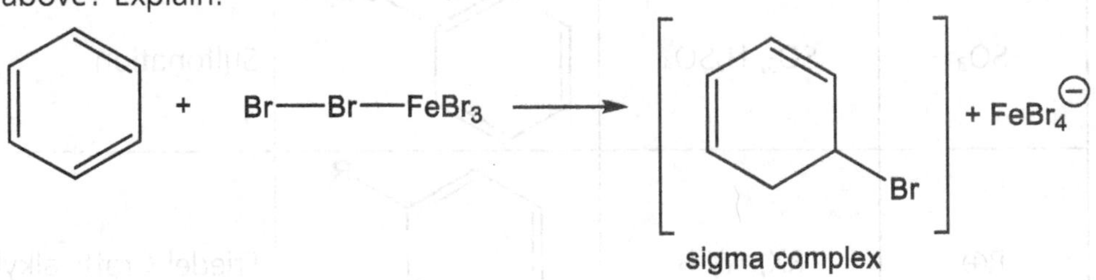

sigma complex

(d)   In the final step of the EAS mechanism, a base removes a proton from the sigma complex to give the substitution product. Draw curved arrows to show which proton is eliminated and to show electron movement in forming the final products. Which species acts as the weak base?

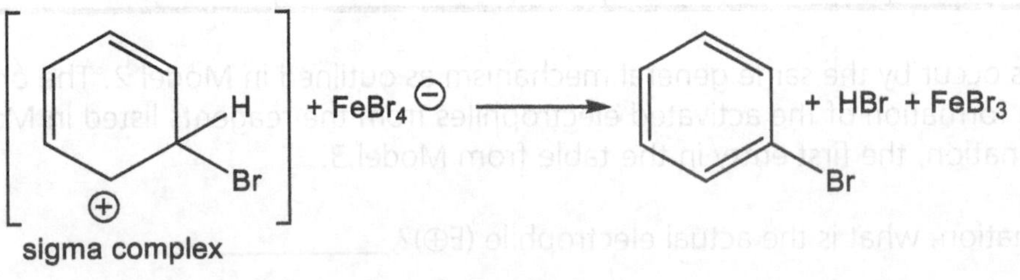

sigma complex

9. Next consider <u>nitration</u>.
   (a)   The nitronium ion, $NO_2^\oplus$, is formed on reaction of nitric acid and sulfuric acid. In the reaction shown, which species acts as the acid? (Hint: look for the conjugate base)

$$HNO_3 \ + H_2SO_4 \longrightarrow \oplus NO_2 \ + HSO_4^\ominus \ + H_2O$$

   (b)   Does the oxygen in the water that is eliminated come from HNO_3 or H_2SO_4?

   (c)   Show how an alcohol is protonated with strong acid. Compare this to protonation of nitric acid by drawing curved arrows to show formation of $NO_2^\oplus$.

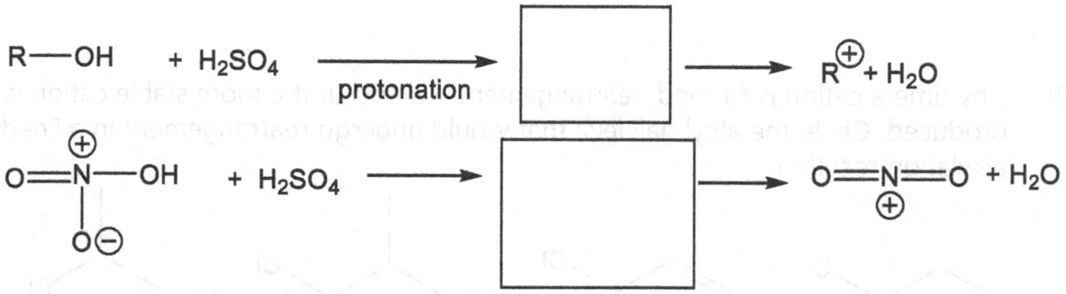

10. Use of a carbon electrophile in EAS reactions is called the <u>Friedel-Crafts reaction</u>.
    (a)   The alkyl chloride is activated by reacting with $AlCl_3$ as shown below. Draw curved arrows to show electron movement in the reaction below. Put charges on the intermediate complex.

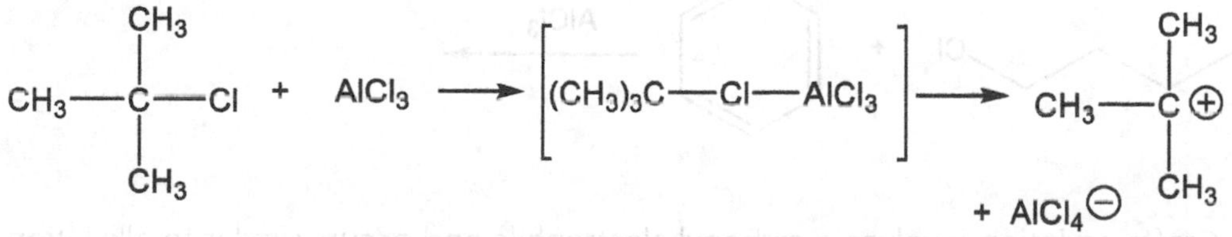

    (b)   Draw the product and the complete mechanism for reaction of benzene with the cation shown above. Make sure to draw all resonance forms of the sigma complex formed.

(c) Secondary and tertiary alkyl halides form the cation prior to reaction with the benzene. As a group discuss whether primary alkyl halides would form the cation. If not, suggest a mechanism for reaction of primary alkyl halides (HINT: think of SN1/SN2 process).

(d) Any time a cation is formed, rearrangements can occur if a more stable cation is produced. Circle the alkyl halide(s) that would undergo rearrangement in a Friedel- Crafts alkylation reaction.

(e) Draw the major organic product of the following reaction.

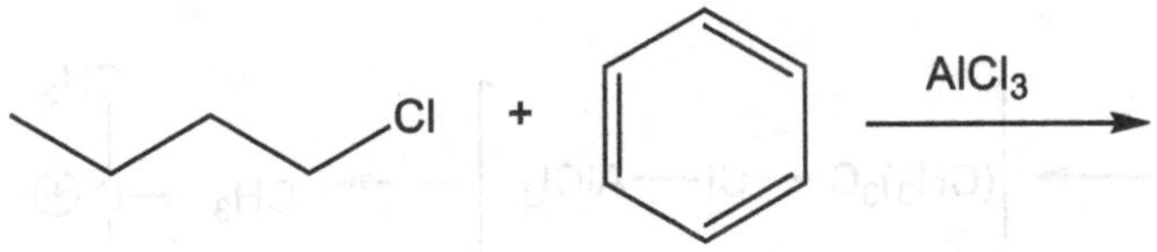

11. Friedel-Crafts acylation involves a carbonyl electrophile and occurs similar to alkylation, to form the cation shown below.

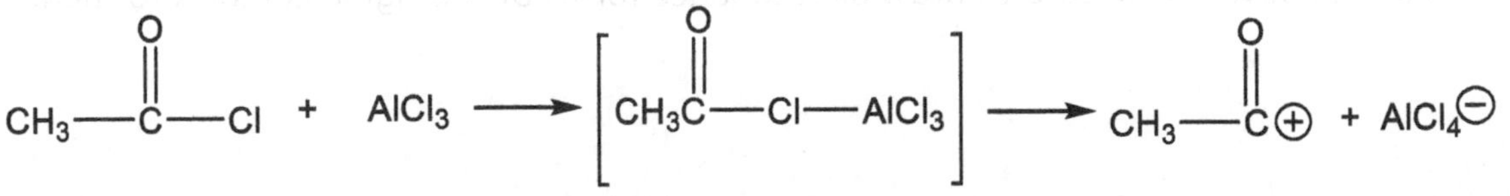

(a) Draw curved arrows to show electron movement in the reaction above. Put charges on the intermediate complex.

(b)   The intermediate cation in Friedel-Crafts acylation does <u>not</u> undergo rearrangement. In your groups discuss reasons why rearrangement will not occur. (HINT: think about what drives rearrangement).

Reflection: on a separate sheet of paper.

As a group, describe three concepts your group has learned from this activity and the one most important unanswered question about this activity that remains with your group. Turn this in before leaving class.

Additional Exercises:

12. Draw the product of the reaction of benzene with each of the following reagents.
    (a) H_2SO_4, HNO_2
    (b) Br_2, $FeBr_3$
    (c) CH_3CH_2Cl, $AlCl_3$
    (d) Cl_2, $AlCl_3$
    (e) CH_3COCl, $AlCl_3$

13. Explain why Friedel-Crafts alkylation cannot be used to form the product shown in the reaction below.

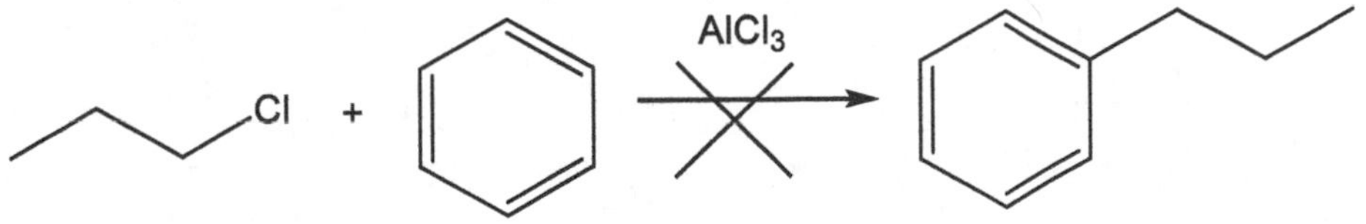

Class Activity 20B

Electrophilic Aromatic Substitution
Part B: Substituent Effects

Prior Knowledge:
Before beginning this activity, students should be familiar with the following concepts:

- Electrophilic aromatic substitution reactions with benzene.
- Energy diagrams, endothermic, exothermic, activation energy.
- Electron donating and withdrawing groups.
- Resonance structures.
- Curved arrow notation.
- Nomenclature of aromatic compounds.

Learning Objectives
Content Learning Objectives:
After completing this activity students should be able to:

- Predict compounds that would be activators and those that would be deactivators.
- Determine whether a compound is an o,p-director or a m-director.
- Articulate why a compound is an o,p- or m-director based on intermediate stability.

Process Objectives:

- Information Processing. Students manipulate resonance structures, intermediate stability and energy diagrams.
- Critical Thinking. Students evaluate directive effects of various substituted benzenes to predict the regioselectivity of addition.

Class Activity 20B

Electrophilic Aromatic Substitution
Part B: Substituent Effects

Model 1: Rate Effects of Substituted Benzenes

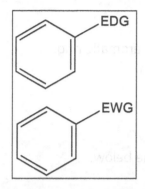

EDG= electron donating group

EWG= electron withdrawing group

Questions:

1. Based on information shown in Model 1:

   (a) An EDG (*circle one*) donates / withdraws electrons to/from the benzene ring. Draw an arrow to show the polarity of the bond between the benzene ring and the EDG.

   The EDG makes the aromatic ring (*circle one*) more / less reactive as a nucleophile.

   (b) An EWG (*circle one*) donates / withdraws electrons to/from the benzene ring. Draw an arrow to show the polarity of the bond between the benzene ring and the EWG.

   The EWG makes the aromatic ring (*circle one*) more / less reactive as a nucleophile.

   (c) An aromatic compound that is <u>more</u> reactive than benzene is called an <u>activator</u>. Which group would make the compound an activator? (*Circle one*) EDG / EWG

   (d) An aromatic compound that is <u>less</u> reactive than benzene is called a <u>deactivator</u>. Which group would make the compound a deactivator? (*Circle one*) EDG / EWG.

2. In your groups discuss what functional groups would be classified as EDGs and EWGs. List at least two groups from each class.

3. (a) Drawing resonance structures can help predict whether a group is an EDG or EWG. Using curved arrows draw three resonance forms of anisole below (HINT: draw the lone pairs of electrons on the oxygen atom first).

(b) The oxygen (*circle one*) donates / withdraws electrons from the aromatic ring.

(c) Anisole would be a(n) (*circle one*) activator / deactivator. Explain.

4. (a) Using curved arrows draw three resonance forms of nitro benzene below.

(b) The nitro group (*circle one*) donates / withdraws electrons from the aromatic ring.

(c) Nitro benzene would be a(n) (*circle one*) activator / deactivator. Explain.

Model 2: Directive Effects of Methylbenzene (toluene)

Questions:

5. Alkyl groups are slightly electron donating. Toluene is a(n) (*circle one*) activator / deactivator.

6. In the nitration of toluene shown in Model 2:
    (a)   What product is formed in Rxn 1? (*Circle one*) ortho / meta / para.
    (b)   What product is formed in Rxn 2? (*Circle one*) ortho / meta / para.

7. (a)   Draw curved arrows in both Rxn 1 and Rxn 2 above to show electron movement.

    (b)   Draw any additional resonance forms for the cationic intermediate (sigma complex) formed in each of the above reactions (one resonance form for each is redrawn below).

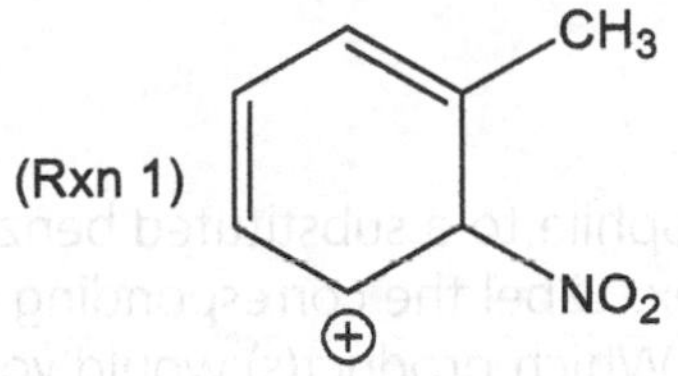

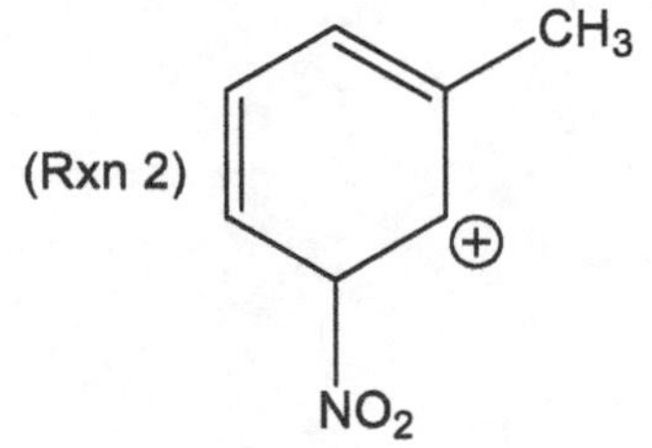

    (c)   Circle any forms that would be a major resonance contributor in Rxn 1 and Rxn 2. As a group discuss what factors make any circled forms a major structure.

    (d)   Of the two products in Model 2, one product forms and the other does not. Circle the product that forms and explain your reasoning.

8. (a) Draw the three resonance forms of the cationic intermediate (sigma complex) that forms when the nitronium ion adds to the <u>para</u> position, as shown below.

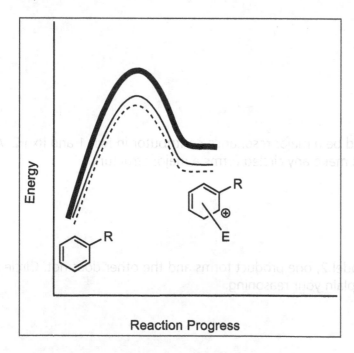

(b) The para-cationic intermediate is very close in potential energy to the ortho-intermediate from Rxn 1 above. In what way are these two sets of resonance structures similar?

(c) The energy diagram below shows addition of an electrophile to a substituted benzene, to form the ortho, meta or para substituted sigma complex. Label the corresponding line on the energy diagram for ortho, meta or para formation. Which product(s) would you expect to be favored? Would you expect all activators to direct to the same positions? Explain.

Model 3: General Directive Effects

(Rxn 3)

(Rxn 4)

Questions:

9. (a)   In Rxn. 3, what products are formed with EDG substituents? (Circle) o / m / p.
    (b)   In Rxn. 4, what product is formed with EWG substituents? (Circle) o / m / p.
    (c)   What groups (EDG or EWG) would be o,p-directors? What groups would be m-directors?

10. Consider the reaction of anisole with an electrophile.
    (a)   Anisole would be a(n) (circle one) activator / deactivator.
    (b)   Draw the three additional resonance forms for the cationic intermediate (sigma complex) when the ortho product is formed.

sigma complex:
draw three additional forms

    (c)   Repeat the above process for the sigma complex formed as a result of meta addition.

(d)   What is the main difference between the resonance forms drawn for the <u>ortho</u> product compared to the resonance forms drawn for the <u>meta</u> product?

(e)   Would you expect there to be any difference in <u>energy</u> between the sigma complex in the ortho product and the sigma complex of the meta product? Explain.

(f)   Would you expect there to be a major difference in energy between the sigma complex for the ortho complex as compared to the para complex? Explain.

OCH₃

+ E⊕ ⟶

11. Usually an activator is an ortho, para-director, and a deactivator is a meta-director. However, a halogenated benzene is a deactivator and ortho-, para-director.

(a)   Draw bromobenzene. Place a polarity arrow on the C-Br bond to indicate the direction of electron flow. Using this information determine what effect induction would have on the rate of bromobenzene in an EAS (increase or decrease rate).

(b)   Draw the sigma complex formed on reaction of bromobenzene with an electrophile. Draw any resonance forms. What effect would resonance have on the directive effect (o,p vs m)?

12. Consider the reaction of nitrobenzene with an electrophile.
    (a)   Nitrobenzene would be a(n) (circle one) activator / deactivator.

    (b)   Draw the additional resonance forms for the cationic intermediate (sigma complex) for the ortho product and the meta product.

    (c)   In your groups discuss any differences in energy between the sigma complexes for ortho and meta. Explain how these differences lead to a preference for the meta product with EWG.

Model 4: Steric Effects - ortho vs para

	ortho	meta	para
R = CH$_3$	40%	3%	57%
R = iPr	12%	3%	85%

Questions:

13. Consider the nitration of an alkyl benzene as shown in Model 4.

    (a) When R is an alkyl group, what nitration products would be formed preferentially? Explain.

    (b) In Model 4, compare differences in product distribution when R=CH$_3$ versus R=iPr.

    (c) Explain why there is a strong preference for the para product when R =iPr rather than R=CH$_3$.

Reflection: on a separate sheet of paper.

As a group, describe three concepts your group has learned from this activity and the one most important unanswered question about this activity that remains with your group. Turn this in before leaving class.

Additional Questions:

14. Draw the major organic products in each of the following reactions.

(a)

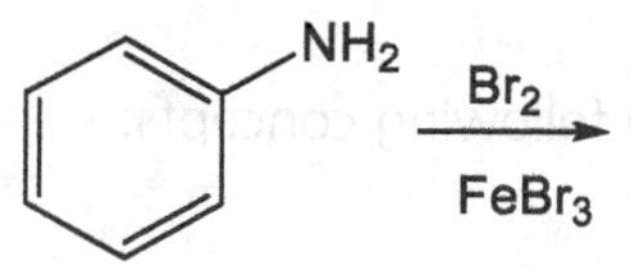

$$Ar\text{-}NH_2 \xrightarrow[\text{FeBr}_3]{\text{Br}_2}$$

(b)

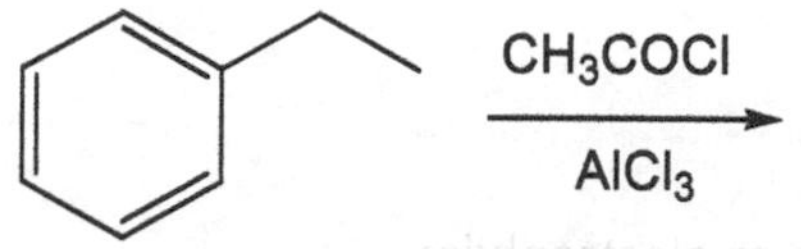

$$\xrightarrow[\text{AlCl}_3]{\text{CH}_3\text{COCl}}$$

(c)

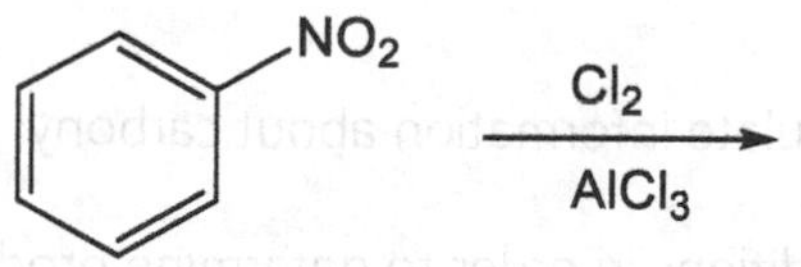

$$\xrightarrow[\text{AlCl}_3]{\text{Cl}_2}$$

(d)

$$\xrightarrow[\text{H}_2\text{SO}_4]{\text{SO}_3}$$

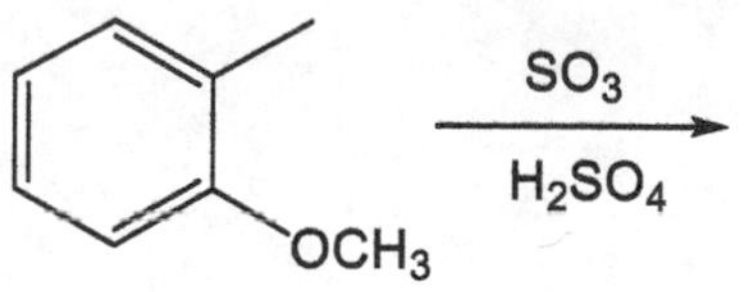

15. Draw the mechanisms for each of the above reactions.

Class Activity 21

Reactions of Carbonyls: Aldehydes and Ketones

Prior Knowledge:
Before beginning this activity, students should be familiar with the following concepts:

- Acid, base, nucleophile and electrophile roles and properties.
- Curved arrow notation.
- Ketone and aldehyde nomenclature.
- Grignard reagent.

Learning Objectives
Content Learning Objectives:
After completing this activity students should be able to:
- Predict the relative reactivity of aldehydes and ketones as electrophiles.
- Draw the mechanism of nucleophilic addition to carbonyls under both acidic and basic conditions.
- Draw the products of nucleophilic addition to aldehydes and ketones using a variety of nucleophiles.

Process Objectives:
- Information Processing. Students interpret and manipulate information about carbonyl reactivity.
- Critical Thinking. Students evaluate reagents and conditions in order to determine products and mechanisms of nucleophilic addition to carbonyl compounds.

Class Activity 21

Reactions of Carbonyls: Aldehydes and Ketones

Model 1: Nucleophilic Addition to Carbonyls

Questions:

1. For the reaction shown in Model 1:
   (a) A carbonyl (aldehyde or ketone) is reacted with a nucleophile. What is the role of the carbonyl? (*Circle one*) nucleophile / electrophile / acid / base.
   (b) Draw a polarity arrow on the C=O bond to show the direction of electron polarization. In the space above, draw a resonance form of the starting carbonyl compound.
   (c) Which atom of the carbonyl is likely to act as the electrophile? Explain why.

   (d) Draw curved arrows to show electron movement for the reaction in Model 1.

2. (a) Draw 2-propanone and ethanal below. Once your group agrees on the structures of these compounds, note any differences between these two compounds.

   (b) Alkyl groups donate electrons through induction. Based on these electronic effects, which compound would be more reactive as an electrophile? Explain.

   (c) Which group, CH_3 or H, is bigger? Based on steric effects, which compound would be more reactive as an electrophile? Explain.

3. Draw structures for the following compounds and indicate which is most reactive in a nucleophilic addition reaction (rank from most to least reactive). Explain why (two reasons).

    pentanal   3-methyl-2-pentanone   diisopropyl ketone

Model 2: Reactions of Carbonyls under Basic Conditions

Questions:

4. (a) In step 1 of Model 2 are any bonds to H formed or broken? (*Circle one*) yes / no.

    The Grignard reagent acts as a (*circle one*) acid / base / nucleophile / electrophile.

    (b) The Grignard reagent would be a (*circle one*) strong / weak reagent (from #4a).

    (c) Draw curved arrows to show electron movement for step 1.

    (d) Explain why this reaction is considered to be basic, when acid is used in step 2.

5. (a) Draw the product expected upon reaction of NaCN with the ketone shown in Model 2.

    (b) What reagent acts as the nucleophile? (*Circle one*) NaCN / ketone.

    (c) Compare the reaction of NaCN to the reaction of RMgBr, shown in Model 2. As a group discuss what is different and what is the same about these reactions

Model 3: Reactions of Carbonyls under Acidic Conditions

activated electrophile

Questions:

6. For the reaction shown in Model 3:

   (a)   What is the functional group of the starting material? _____

        What is the functional group of the product? _____

        Compare the starting material and the product. What reagent has been added to the carbonyl?

   (b)   Water is added to the carbonyl in the second step. What role does water play? (*Circle one*) nucleophile / electrophile. Would water be considered a strong or a weak reagent? Explain.

   (c)   In the first step, acid is added to the carbonyl. What is this step called? _____

   (d)   Draw a resonance form for the protonated carbonyl. Would you expect the protonated carbonyl to be (*circle one*) more / less reactive as an electrophile? Use the resonance structure to explain the purpose of the protonation step.

   (e)   Draw curved arrows to show electron movement for the remainder of the reaction show in Model 3. Once your group agrees on the mechanism for this hydration reaction, discuss why it is possible to use water as a nucleophile in this reaction.

Model 4: Nucleophiles in Nucleophilic Addition to Carbonyls

A summary of nucleophilic addition reactions to carbonyls is shown below.

Reactant	Nu⊖	Conditions	Product	Reaction Name	Entry
![carbonyl R-CO-R, R = alkyl or H]	H_2O	acidic	R—C(OH)(OH)—R	Hydration	1
	HO⊖	basic	R—C(OH)(OH)—R	Hydration	2
	R'OH	acidic	R—C(OR')(OR')—R	Acetal formation	3
	⊖CN	basic	R—C(OH)(CN)—R	Cyanohydrin formation	4
	1° R'NH₂	acidic	R₂C=NR'	Imine formation	5
	Ph₃P⊕—⊖CH₂	basic	R₂C=CH₂	Wittig Reaction	6

Questions:

7. (a)   Circle the nucleophiles in the table above that would be <u>weak</u> nucleophiles.
    (b)   When reacting a weak nucleophile with a carbonyl compound, an acid catalyst is used. Do the reactions under <u>acidic conditions</u> in the table match the nucleophiles you chose as weak?

What is the purpose of the acid catalyst?

8. (a)   In the acid catalyzed hydration (entry 1), what is the nucleophile? _____
    (b)   In the base catalyzed hydration (entry 2), what is the nucleophile? _____
    (c)   Compare the products from these two hydration reactions. These hydration products are (*circle one*) the same / different.

(d)   Draw the products of the following hydration reactions.

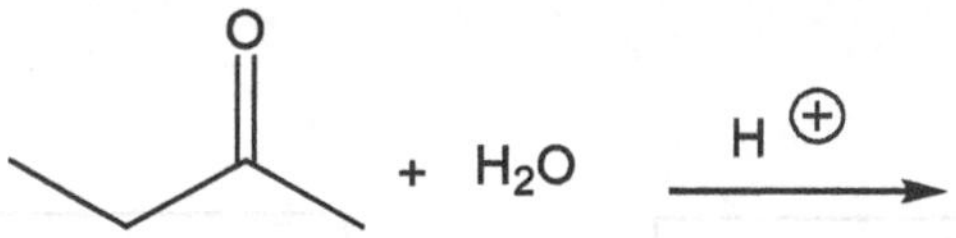

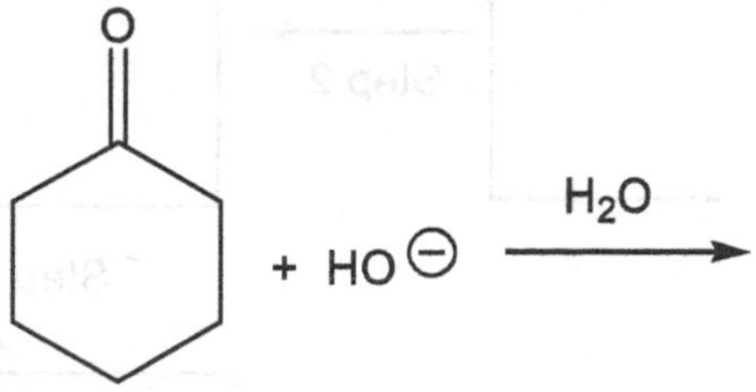

9.   Refer to entry 3 in Model 4 (acetal formation).
   (a)   Acetal formation is conducted under (*circle one*) acidic / basic conditions.
   (b)   In acetal formation what is the nucleophile? _____
   (c)   Consider the acetal product formed in entry 3. Describe the functional groups that make up an acetal.

   (d)   Circle any acetals in the list of compounds shown below.

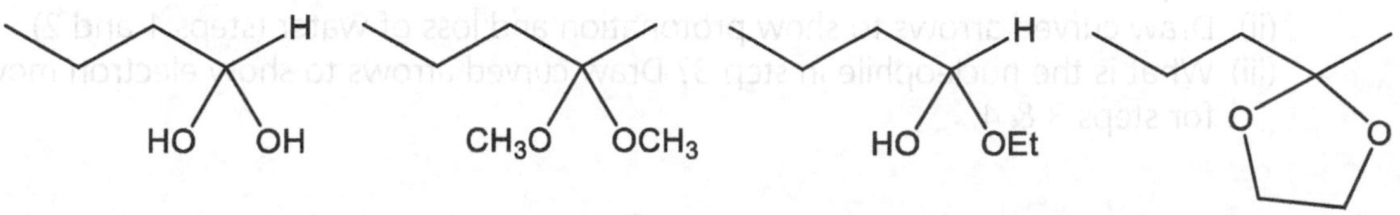

   (e)   Acetal formation is similar to the acid catalyzed hydration in Model 3. Draw the mechanism for acid catalyzed acetal formation after one equivalent of alcohol has been added to give the hemiacetal (shown below). Use curved arrows to show electron movement.

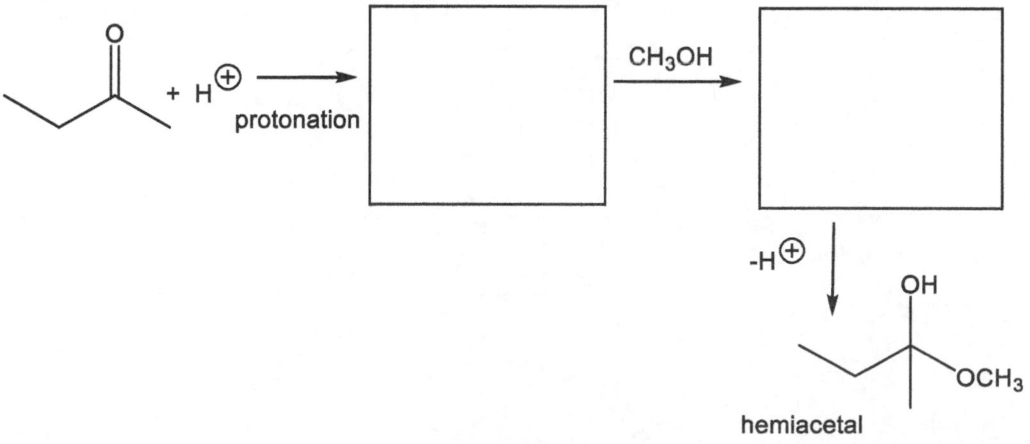

(f)   The steps from hemiacetal to acetal are similar to protonation and substitution of an alcohol.

(i)   Circle the group that must be protonated in step 1 in order to eliminate water in step 2.

(ii) Draw curved arrows to show protonation and loss of water (steps 1 and 2).

(iii) What is the nucleophile in step 3? Draw curved arrows to show electron movement for steps 3 & 4.

(g)   Once your group has agreed on the above mechanism, predict the reagents needed to make the acetals circled in question #9d above.

10. Consider the entry 6 in the table above, the Wittig reaction.

(a) The nucleophile in a Wittig reaction is the phosphorus ylide, below. Circle the atom of the phosphorus ylide that reacts as the nucleophile. Draw an additional resonance form.

$$Ph_3\overset{\oplus}{P}\!\!-\!\!\overset{\ominus}{\ddot{C}H_2}$$

(b) Draw the <u>first</u> step of the reaction of the phosphorus ylide with dimethylketone (acetone). Use curved arrows to show electron movement (HINT: a carbon-carbon bond is formed).

(c) Given that phosphorus oxygen bonds are strong (and readily formed), draw the four membered ring product formed from #10b above.

(d) The final products are an alkene and $Ph_3P=O$. Draw curved arrows from the four membered ring intermediate shown in #10c to explain how these products are formed.

Reflection: on a separate sheet of paper.

As a group, describe three concepts your group has learned from this activity and the one most important unanswered question about this activity that remains with your group. Turn this in before leaving class.

Additional Questions:

11. Use curved arrows to predict the complete mechanism of formation of the imine shown below.

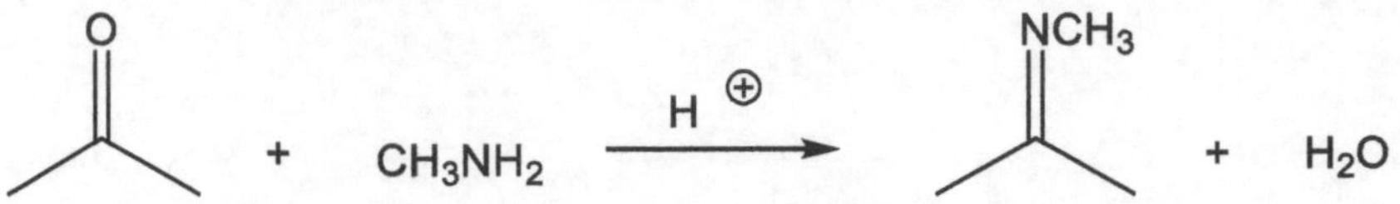

12. Draw the major organic product for each of the following reactions.

(a)

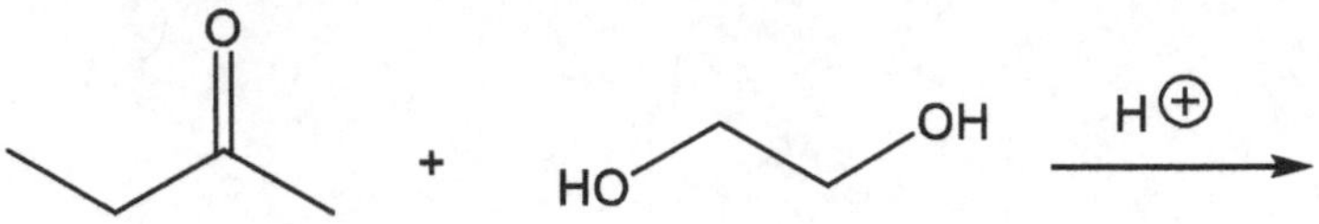

(b)

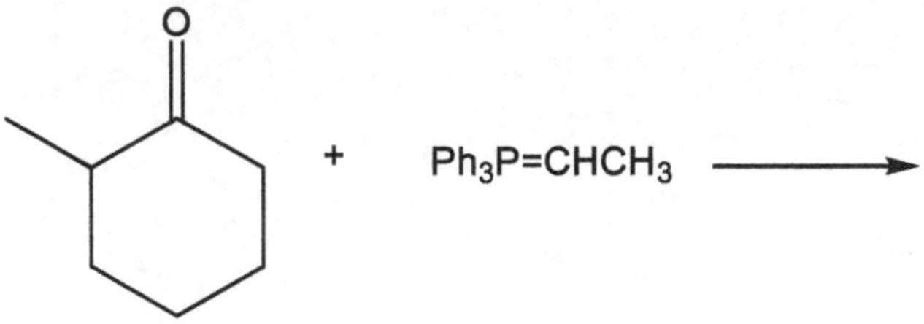

(c)

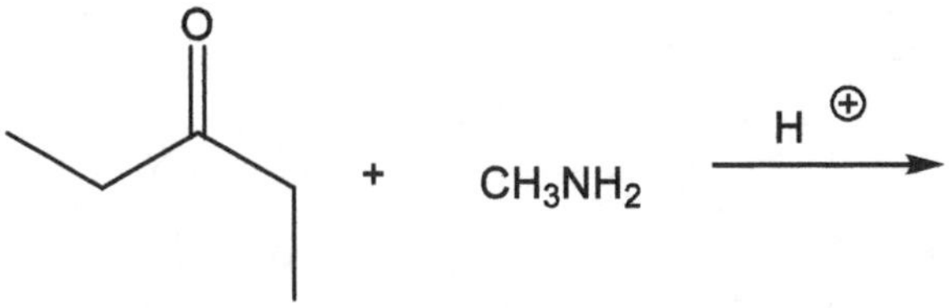

(d)

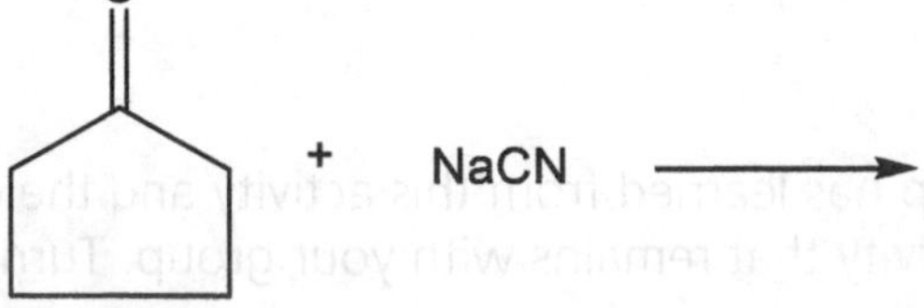

13. Draw starting materials needed to prepare the following compounds.

(a)

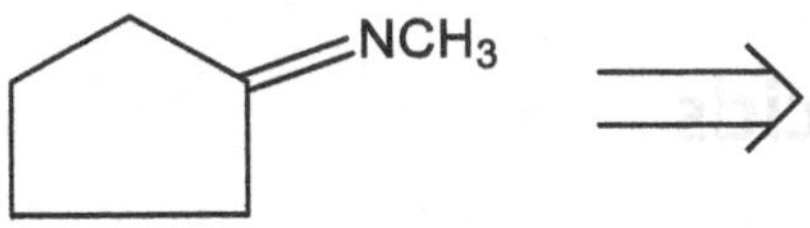

(b)

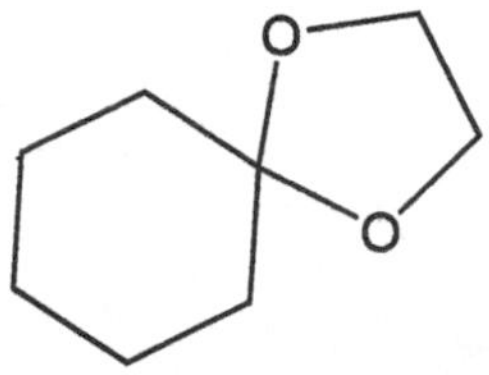

(c)

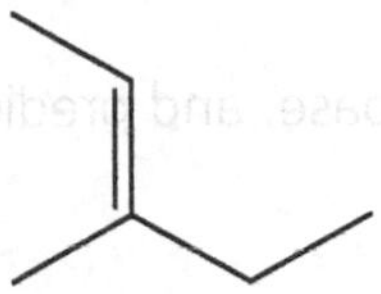

(d)

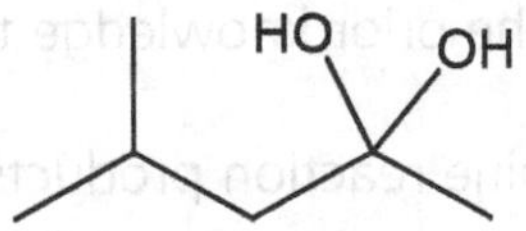

14. Safrole, shown below, is found in the sassafras plant and was used to add flavor to root beer. Circle the acetal functional group and draw the carbonyl and any other starting materials necessary to prepare this acetal.

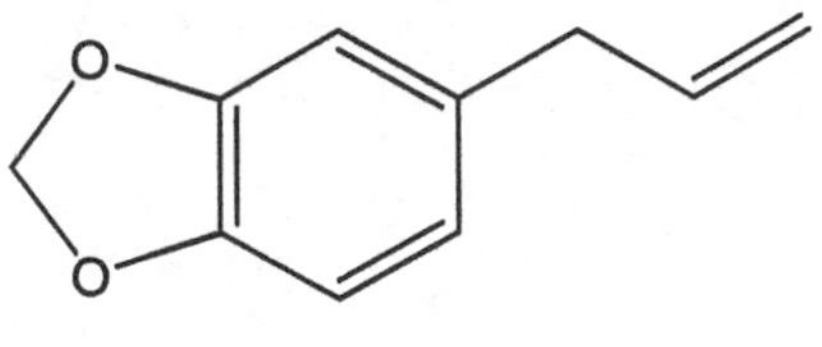

Class Activity 22

Carboxylic Acids

Prior Knowledge:
Before beginning this activity, students should be familiar with the following concepts:

- Acid, base, nucleophile and electrophile roles and properties.
- Factors that stabilize a conjugate base.
- Resonance.
- Curved arrow notation.
- Keto-enol tautomerization.

Learning Objectives
Content Learning Objectives:
After completing this activity students should be able to:
- Determine the roles of a carboxylic acid when reacted with 2eq of base, and predict the resulting ketone product.
- Formulate a mechanism for the Fischer esterification reaction.
- Predict which carboxylic acids will undergo decarboxylation and draw the products.

Process Objectives:
- Information Processing. Students interpret information from the prior knowledge topics as applied to carboxylic acids.
- Critical Thinking. Students evaluate carboxylic acids to determine reaction products under different conditions.

Class Activity 22

Carboxylic Acids

Model 1: Acidity of Carboxylic Acids

Review the effects that stabilize a conjugate base:
1. Size
2. Formal charge
3. Electronegativity
4. Resonance
5. Induction

Questions:

1. Consider the acid/base reaction shown in Model 1.

   (a) For the starting materials on the left, which compound acts as the acid? _____ Which compound acts as the base? _____ Label the starting materials above as acid and base. Using the pKa values, justify your choices.

   (b) For the products on the right, label as either the conjugate base or conjugate acid.

   (c) Compare the base and the conjugate base. Circle the base that is the most stable. Do the pKa values agree with your choice? Explain using the effects that stabilize bases.

   (d) This reaction will proceed in the (*circle one*) forward / reverse direction.

2. Draw the conjugate base for each carboxylic acid below. Determine which conjugate base is more stable and justify using the relevant factors reviewed in Model 1. Circle the acid that would have the lowest pKa.

3. Rank the following acids from most acidic to least acidic. Explain the ranking using the effects that lead to stabilization of the conjugate base.

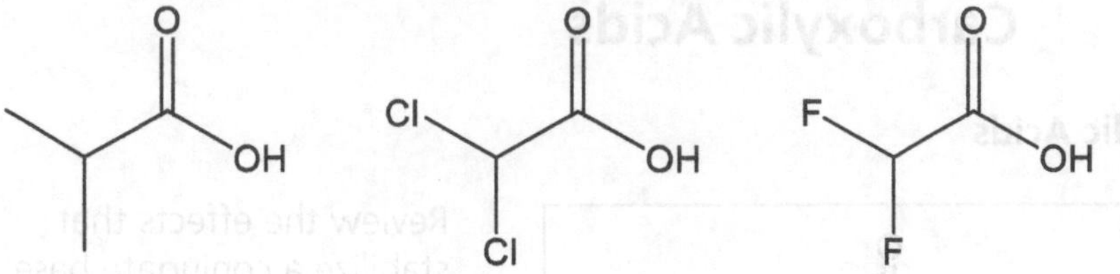

Model 2: Alkylation of Carboxylic Acids

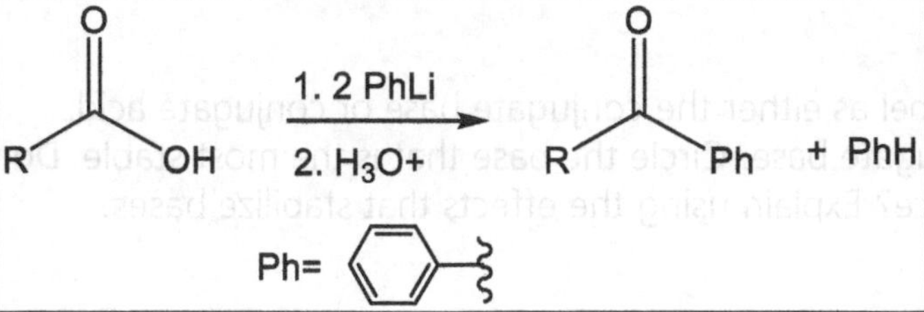

Questions:

4. For the reaction shown in Model 2:

(a) What is the functional group of the starting material?_____

What is the functional group of the product? _____

What group has been added to the starting material to make the product? _____

What reagent did this group originate from? _____

(b) PhLi has a pKa = 50. When reacted with a carboxylic acid, PhLi will initially act as a (*circle one*) Nu⊖ / E⊕ / acid / base.

(c) Draw the product after the first equivalent of PhLi is added to the carboxylic acid. (HINT: acid/base reaction). Draw all resonance forms of the resulting conjugate base.

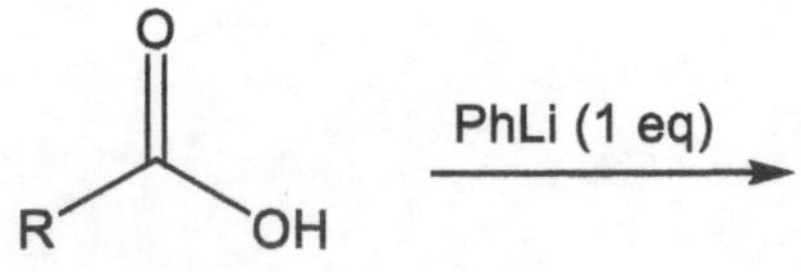

(d)   Consider the resonance forms of the conjugate base in #4c. Circle the <u>atom</u> that contains a positive charge in one of the resonance forms. If reacted with a Nu⊖, this atom will act as the (*circle one*) Nu⊖ / E⊕ / acid / base. Draw the product expected when the conjugate base from #4c reacts with the second equivalent of PhLi.

(e)   In your groups discuss why it is essential to add <u>two</u> equivalents of alkyl lithium in the alkylation of a carboxylic acid.

5. The following compound is formed after two equivalents of PhLi are added to the carboxylic acid. Does this agree with your answer to #4d? Draw the neutral product that would be formed after addition of excess H_3O+ in step 2.

6. The neutral product from #5 reacts under acidic conditions to give the final product from Model 2, with loss of water. Propose a mechanism for this step. (HINT: the neutral product is an alcohol, reacting under acidic conditions).

Model 3: Fischer Esterification

tetrahedral
intermediate

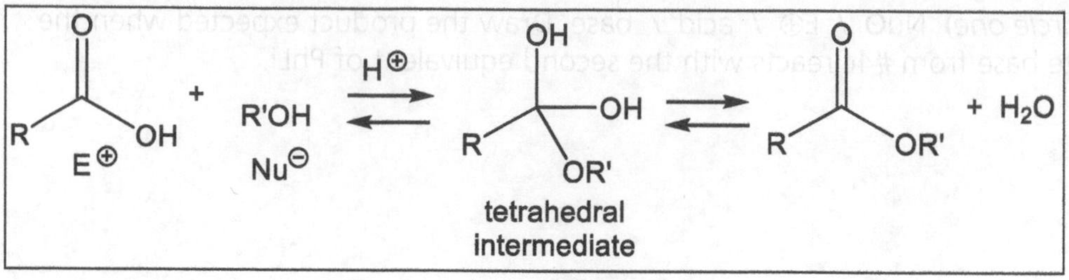

Questions:

7. For the Fischer Esterification reaction shown in Model 3:
   (a) Which compound acts as the electrophile? _____
       This electrophile is (*circle one*) strong / weak. Explain.

       Which compound acts as the nucleophile? _____.
       This nucleophile is (*circle one*) strong / weak. Explain.

   (b) What is the purpose of the acid catalyst, H⊕?

8. Consider the reaction of a <u>ketone</u> and an alcohol under acidic conditions to form an acetal (from Class Activity 21).

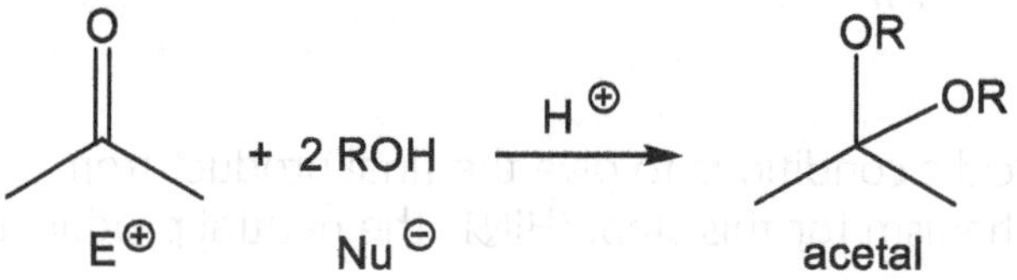

acetal

   (a) The alcohol acts as the nucleophile. This nucleophile is (*circle one*) strong / weak.
   (b) Draw the <u>first</u> step of this mechanism, when the ketone reacts with H⊕.

(c)   What is the purpose of the acid catalyst, $H\oplus$?

(d)   Does your answer to #8c agree with your answer to #7b? What is similar about the reaction of $H\oplus$ with a ketone compared to the reaction of $H\oplus$ with a carboxylic acid?

9. (a)   Draw the product obtained after the first step in the reaction of carboxylic acid with $H\oplus$ shown in Model 3. (HINT: it should be similar to the reaction of the ketone in #8b). Draw all possible resonance forms of this compound.

(b)   The protonated carboxylic acid from #9a would be (*circle one*)   more / less reactive as an electrophile, than the neutral carboxylic acid.

(c)   Once protonated, the carboxylic acid can react with the neutral alcohol to form the tetrahedral intermediate (see Model 3). Draw this process, using curved arrows to show electron movement.

(d)   Why is the intermediate called the tetrahedral intermediate?

10. Consider the second step of the Fischer esterification shown in Model 3.
    (a)   What molecule is removed from the tetrahedral intermediate to form the ester?

    (b)   The tetrahedral intermediate is an alcohol which undergoes acid catalyzed dehydration. Propose a mechanism for this step, using curved arrows to show electron movement.

11. The Fischer esterification is an equilibrium reaction. In your groups, discuss how this reaction could be forced to the right, to form the ester.

Model 4: Decarboxylation of Carboxylic Acids

Questions:

12. For the reactions shown in Model 4:
    (a)   What molecule has been eliminated after heating the carboxylic acid in Eq. 1?
    (b)   After heating the carboxylic acid shown in Eq. 2, has any reaction occurred?
    (c)   Number the carboxylic acids in Eq. 1 and Eq. 2, and provide an IUPAC name for each.

    (d)   The ketone group occurs at carbon number _____ for the keto-acid in Eq. 1, and at carbon number _____ for the keto-acid in Eq. 2. Based on these observations, discuss the structural requirements in order for decarboxylation to occur.

(e) Alternatively, Greek letters can be used to indicate the position of the carbonyl group with respect to the carboxylic acid. Carbon #2 of a carboxylic acid is the α–carbon. Label the carbons above as α, β, and γ. At what position must the carbonyl group be in order for decarboxylation to occur? (*Circle one*) α / β / γ.

(f) Circle the compounds of those shown below that would be capable of undergoing decarboxylation upon heating. For those structures that are not circled, provide an explanation for why they cannot undergo decarboxylation.

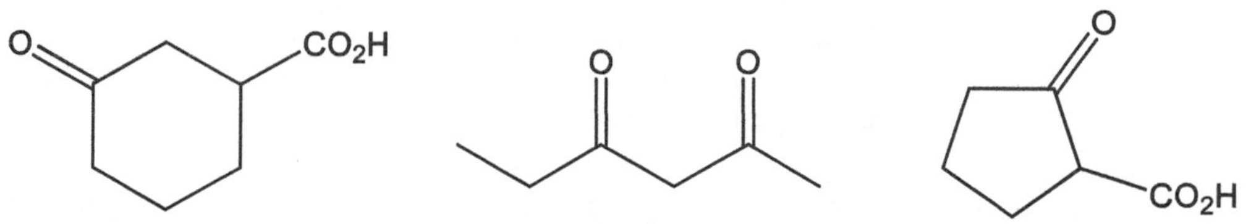

13. Decarboxylation occurs because of the ability to form a hydrogen bond between the carboxylic acid and the neighboring carbonyl group, as shown below.

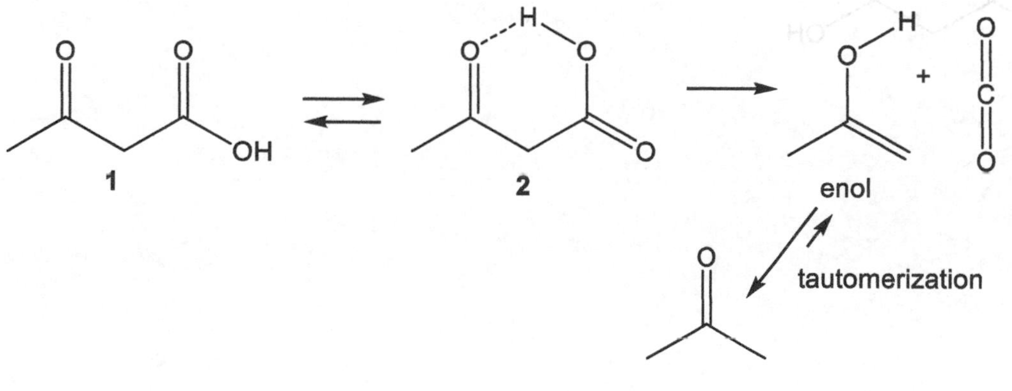

(a) Number the carbons of the carboxylic acid in structures **1** and **2** shown above (the carboxylic acid carbon is #1). At what position is the ketone carbonyl? ___

(b) What is the relationship between structures **1** and **2**? (*Circle one*) constitutional isomers / conformational isomers / not isomers.

(c) The hydrogen bond is drawn as a dotted line in structure **2**. Draw curved arrows to show the electron movement to allow loss of CO_2 to give the enol. (There should be three arrows drawn).

(d) Decarboxylation occurs only when a cyclic six-membered transition state can form through the hydrogen bonding effects as shown above. Explain why each of the following compounds do <u>not</u> undergo decarboxylation.

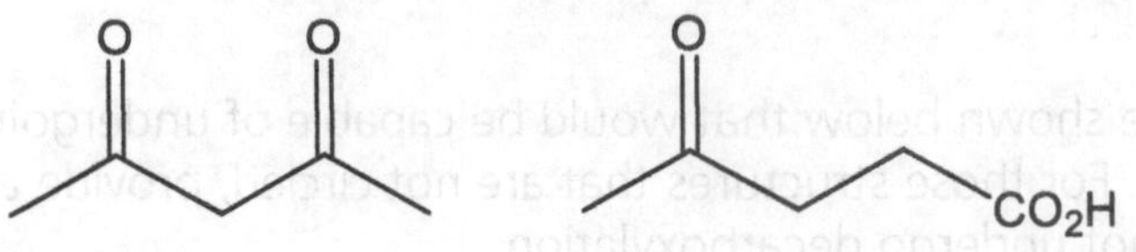

(e) Consider the di-carboxylic acid shown below. Would this compound be capable of undergoing decarboxylation on heating? Explain.

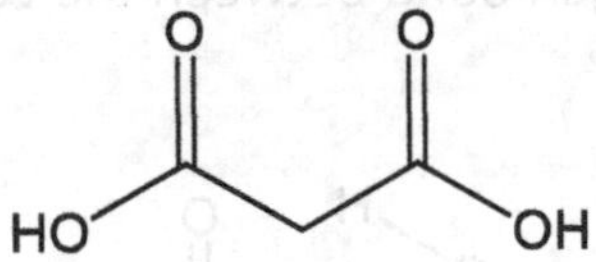

Reflection: on a separate sheet of paper.
As a group, describe three concepts your group has learned from this activity and the one most important unanswered question about this activity that remains with your group. Turn this in before leaving class.

Additional Questions:
14. Circle the most acidic compound in each of the following pairs. Explain your choice by describing the relevant factors for stabilizing the conjugate base.

(a)

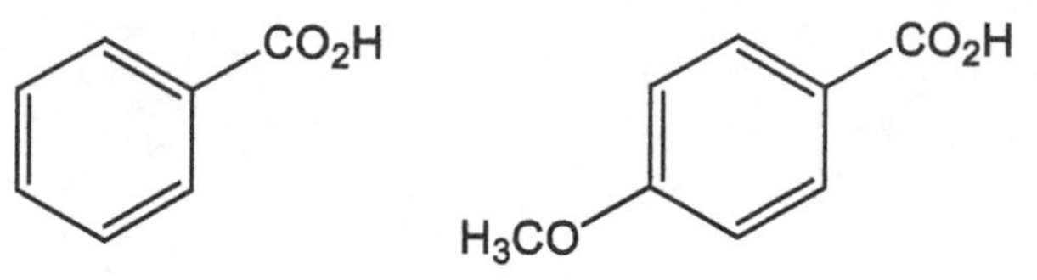

(b)

(c)

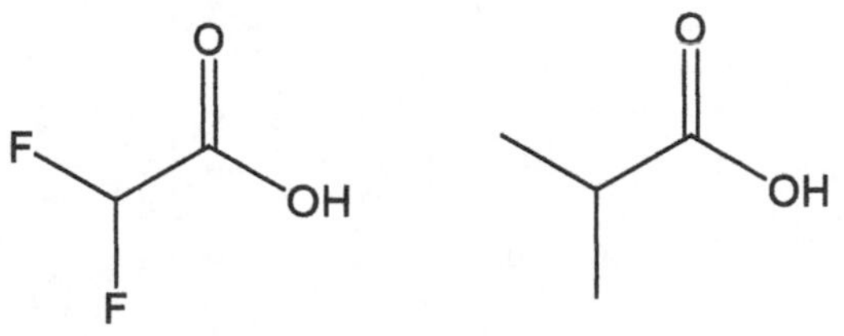

15. Draw the major organic product formed for each of the following reactions.

(a)

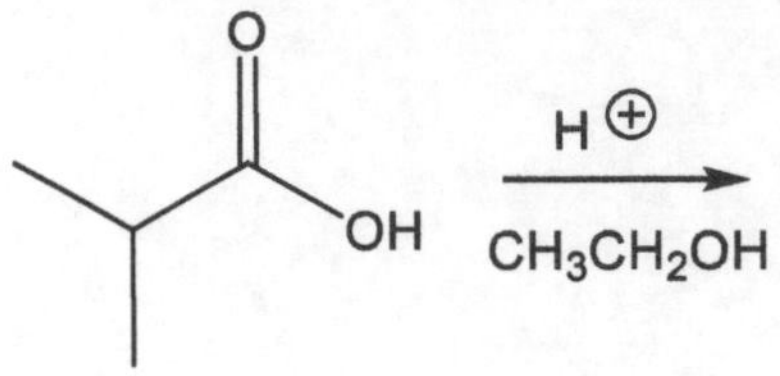
O
OH
H ⊕
CH3CH2OH

(b)

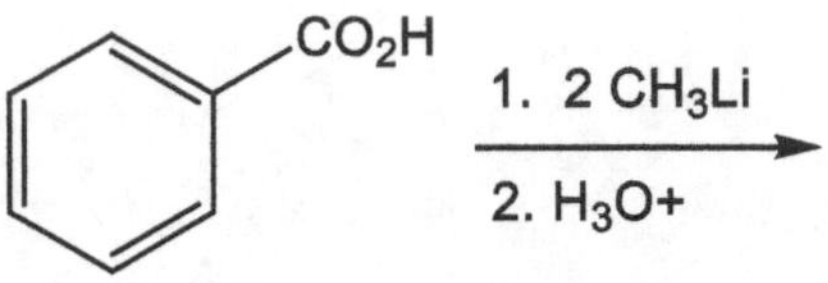
CO2H
1. 2 CH3Li
2. H3O+

(c)

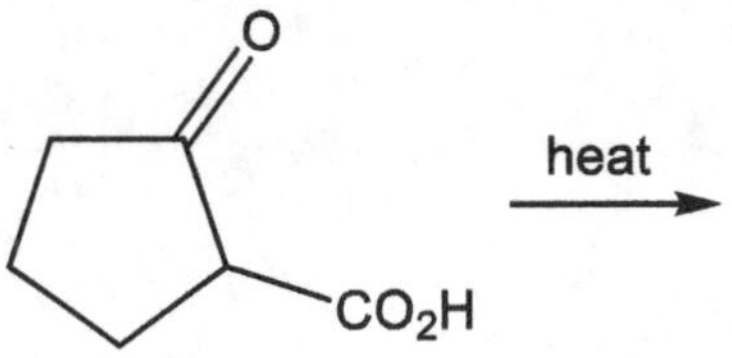
O
heat
CO2H

16. Draw the products formed after each step of the following synthetic sequence.

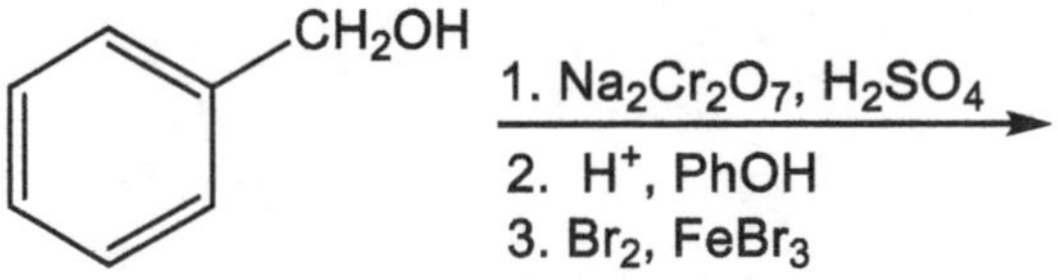
CH2OH
1. Na2Cr2O7, H2SO4
2. H+, PhOH
3. Br2, FeBr3

Class Activity 23A

Acidity at the α–Carbon of Carbonyls

Prior Knowledge:

Before beginning this activity, students should be familiar with the following concepts:

- Acid, base, nucleophile and electrophile roles and properties.
- Formal charge, electronegativity, resonance, induction.
- Curved arrow notation.
- Ketone and aldehyde nomenclature.
- Alkene stability.

Learning Objectives

Content Learning Objectives:

After completing this activity students should be able to:
- Label the positions of a carbonyl (α, β, etc.) and determine which H is the most acidic.
- Distinguish between enols and enolates and determine the more stable form in an unsymmetrical system.
- Predict the product and mechanism for alkylation and halogenation of a carbonyl under basic conditions.

Process Objectives:

- Information Processing and Critical Thinking. Students interpret and analyze concepts of acidity, nucleophilicity, addition to carbonyls and substitution reactions to reach a conclusion about how enolates form and are reacted.

Class Activity 23A

Acidity at the α–Carbon of Carbonyls

Model 1: Acidity of Carbonyl Compounds

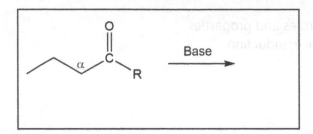

Review the effects that stabilize a conjugate base:
1. Size
2. Formal charge
3. Electronegativity
4. Resonance
5. Induction

Questions:

1. Consider the carbonyl compound in Model 1:
    (a)   What Greek character is given to the carbon directly attached to the carbonyl carbon?
    (b)   How many α carbons are there in an aldehyde (when R=H)? _____
    (c)   How many α carbons are there in a ketone (when R=CH₃)? _____
    (d)   Label all remaining carbon atoms as β, γ, δ etc.
    (e)   Any H attached to the α-carbon, are called α-hydrogens. Draw in the α-hydrogens for the carbonyl shown above.

2. Consider the reaction of an aldehyde (R=H) with a base as shown in Model 1.
    (a)   How many different kinds of H's are there in this aldehyde?

    (b)   Draw the conjugate base that results from removal of the indicated hydrogens below.

    Base
    remove H attached to carbonyl

    Base
    remove α-H

    Base
    remove β-H

(c)   Draw all possible resonance structures of these conjugate bases.

(d)   Which conjugate base from #2b would have the lowest energy? Explain your choice by describing any effects from stabilization of the conjugate base outlined in Model 1.

(e)   Circle the H on the aldehyde that takes the least energy to remove. This is the most acidic H.

3. A base can also act as a nucleophile.
    (a)   A Lewis base (*circle one*) donates / accepts electrons.
    (b)   Describe what is the same between a Lewis base and a nucleophile. What is different?

(c)   Draw the product that would result when the carbonyl reacts with a <u>nucleophile</u>.

(d)   The reaction shown in #3c, competes with the acid/base reaction. In your groups, discuss how the base/nucleophile could be altered in order to force the acid/base reaction to proceed preferentially.

4. (a) Determine whether each of the following compounds would act predominately as a base, a nucleophile or both. Explain your answers.

LDA

(b) Draw the most likely product that forms on reaction of butanal and LDA.

Model 2: Enols and Enolate Ions

keto form   enolate ion

tautomerization   H₂O

:ÖH

enol

Questions:

5. (a)   What is the main difference between the enolate ion and the enol shown in Model 2?

    (b)   Which H is removed to form the enolate ion and the enol? (*Circle one*) Hα / Hβ / Hγ.

    (c)   Is base used to form the enolate ion? (Y or N) Is base used to form the enol? (Y or N)

    (d)   The interconversion between the ketone and enol is an equilibrium reaction. What is this keto-enol interconversion called?_____ Which form is favored (keto or enol)? Explain why.

6. (a)   Draw the enol form of the following ketone.

    (b)   Draw the keto form of the following enol.

7. (a)   Unsymmetrical ketones can form two enol forms. Draw the two enol forms expected from the following ketone (remember to remove the α-H's).

    (b)   Circle the enol above that would be the most stable. (HINT: think of double bond stability). This is called the thermodynamic enol.

8. Consider the ketoaldehyde shown below.
    (a)   Label all the α - carbons.

    (b)   Draw the two possible enolates that could be formed by removing each α–hydrogen on
           reaction with base. Draw all possible resonance forms for both of the enolates

    (c)   Based on the stability of the conjugate base (the enolate) discuss which hydrogen is the
           most acidic in the above ketoaldehyde. Once your group has reached agreement, circle
           the most acidic hydrogen.

Model 3: Reactions α to a Carbonyl

enolate ion

Questions:

9. For the reaction shown in Model 3:
    (a)   Label the α–carbon in the aldehyde structure above.

    (b)   In step one, LDA is combined with the aldehyde to give the enolate ion. Discuss the
           purpose of adding a strong bulky base to the aldehyde.

(c) In step 2, an electrophile is added to the enolate ion. What is the role of the enolate ion in the second step? (*Circle one*) Nu⊖ / E⊕.

(d) Draw the product and the mechanism when CH_3Br is used as the electrophile. Once your group agrees on the mechanism, discuss whether this mechanism supports the roles just assigned (i.e. does the nucleophile donate electrons in your mechanism?).

(e) After formation of the enolate ion, what mechanism is occurring on reaction with the primary alkyl halide? (*Circle one*) SN1 / SN2 / E1 / E2. This reaction is called alkylation. Explain.

10. Predict the products of the following reactions.
    (a)

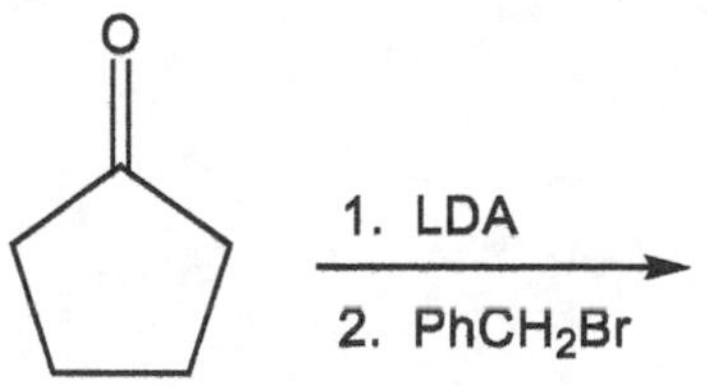

        1. LDA
        2. $PhCH_2Br$

    (b)

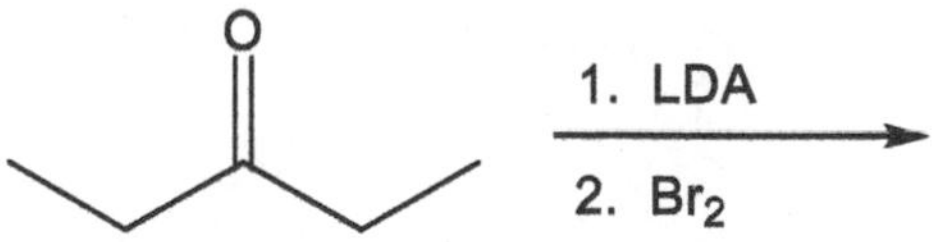

        1. LDA
        2. Br_2

Reflection: on a separate sheet of paper.
   As a group, describe three concepts your group has learned from this activity and the one most important unanswered question about this activity that remains with your group. Turn this in before leaving class.

Additional Questions:

11. Draw phenol and its keto form. Explain why phenol is more stable as the enol than the keto form.

12. Enol formation can also occur under acidic catalyzed conditions. Predict a mechanism for this reaction. (HINT: carbonyl compounds are protonated under acidic conditions).

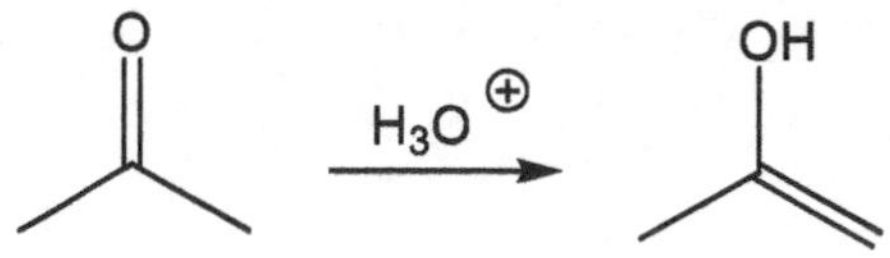

13. Propose a mechanism for the following reaction.

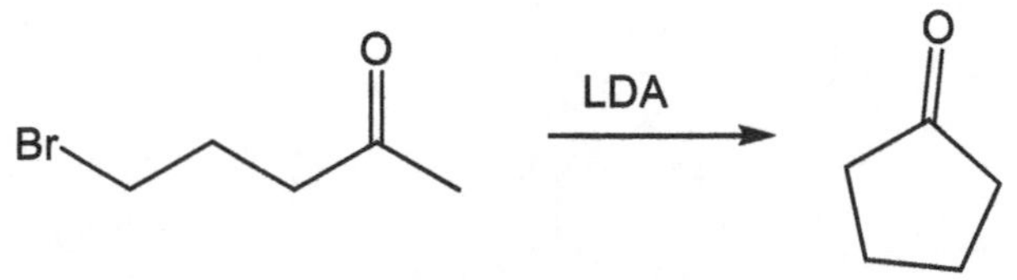

14. Provide reagents that would accomplish the following synthesis.

Class Activity 23B

Aldol Condensations

Prior Knowledge:
Before beginning this activity, students should be familiar with the following concepts:

- Acid, base, nucleophile and electrophile roles and properties.
- Formal charge, electronegativity, resonance, induction.
- Curved arrow notation.
- Ketone and aldehyde nomenclature.
- Enolate formation.

Learning Objectives
Content Learning Objectives:
After completing this activity students should be able to:
- Predict the product and draw the mechanism for the aldol condensation reaction (both self-condensation and mixed).
- Determine the starting materials required when given the aldol product.
- Predict the dehydration product of an aldol condensation and determine the aldol starting materials.

Process Objectives:
- Critical thinking. Students analyze and evaluate the role of a carbonyl as both a nucleophile and an electrophile.
- Problem Solving. Students plan a strategy to determine the mechanism for the aldol self-condensation and mixed condensation reaction.

Class Activity 23B

Aldol Condensations

Model 1: The Aldol Condensation

O
‖   OH O
                                      | ‖
                ⊖ OH (cat.)
      ⇌
        H₂O

propanal   aldol product

Questions:

1. Consider the aldol condensation shown in Model 1.
   (a) How many carbons are in propanal? _____
   (b) How many carbons are in the aldol product? _____
   (c) How many molecules of propanal combine to form the aldol product? _____ Circle each propanal portion in the aldol product.
   (d) Using Greek characters, label the carbons of the product as α, β, γ.
   (e) Using Greek characters, indicate the position where the new C-C bond occurs (two characters).
   (f) At what position does the hydroxyl (OH) group reside? (*Circle one*) α / β / γ / δ
   (g) Name the two functional groups in the aldol product.

   i. _____ ii._____

   Explain why "aldol" is an appropriate name for this compound.

2. In an aldol condensation, two equivalents of an aldehyde (propanal in Model 1) combine to form the aldol product under base catalyzed conditions. One equivalent of aldehyde acts as the electrophile while the other equivalent acts as the nucleophile.

   O   O   OH   O
   ‖   ‖   ⊖OH (cat.) | ‖
             +   ⇌
                     H   H   H₂O   H
   E⊕   Nu ⊖

   (a) Consider the underlined electrophile. Circle the atom that is the most positive in the E⊕ above. Explain.

(b) When an aldehyde reacts with a strong nucleophile as shown below,

$$RCHO + Nu^{\ominus} \rightarrow RCH_2OH$$

the carbonyl group is converted to a(n) _____ (functional group) after aqueous workup.

(c) Circle the atom in the aldol _product_ that came from the carbonyl of the electrophile.

(d) Consider the _nucleophile_. Which carbon of the nucleophile forms a new bond with the electrophile? (_Circle one_) α / β / γ. Draw a box around this carbon in the Nu$\ominus$ and in the product above.

(e) Draw the reaction (using the base catalyst) that must occur in order to convert the aldehyde into a nucleophile at the α–carbon. (HINT: Refer to CA23A).

(f) Once your group agrees on the above questions, discuss why the following aldehydes _cannot_ undergo an aldol condensation reaction.

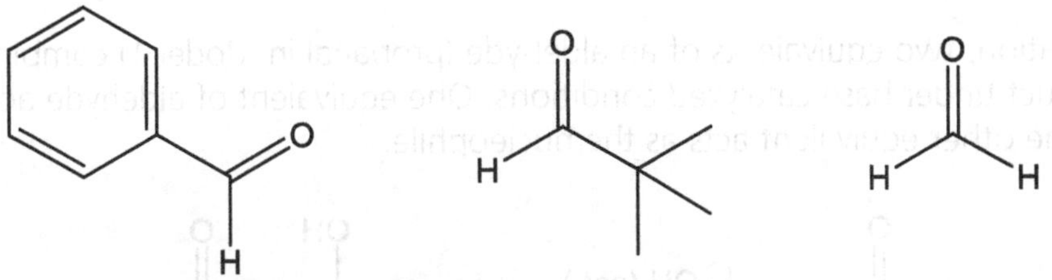

3. Devise a mechanism for the aldol condensation, using curved arrows to show electron movement.

Model 2: Mixed (Crossed) Aldol Condensation

Questions:

4. (a) Provide names for aldehydes A and B in Model 2. Are these two aldehydes the same? (*Circle one*) yes / no.

(b) Label each α-hydrogen on compound A and B in Model 3.

(c) Can compound A act as the Nu$\ominus$? (*Circle one*) yes / no.

    Can compound A act as the E$\oplus$? (*Circle one*) yes / no.

(d) Can compound B act as the Nu$\ominus$? (*Circle one*) yes / no.

    Can compound B act as the E$\oplus$? (*Circle one*) yes / no.

(e) If compound A reacted with another molecule of compound A, this would be called (*circle one*) self / mixed condensation.

(f) If compound A reacted with a molecule of compound B, this would be called (*circle one*) self / mixed condensation.

(g) In the boxes below, draw the two self-condensation products and two mixed condensation products that result from the mixture in Model 2. Use the roles (Nu⊖, E⊕) for each reagent listed in each box to determine the products.

Self Condensation Nu⊖ = A E⊕ = A	Self Condensation Nu⊖ = B E⊕ = B
Mixed Condensation Nu⊖ = A E⊕ = B	Mixed Condensation Nu⊖ = B E⊕ = A

5. Mixed condensations can also occur between ketones and aldehydes.
   (a) Which of the following would be the stronger electrophile? (*Circle one*) aldehyde / ketone. Explain.

   (b) In the following reaction explain why benzaldehyde cannot act as a nucleophile.

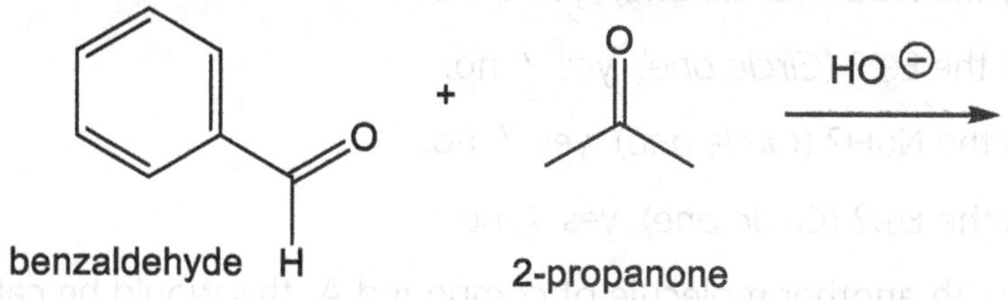

benzaldehyde H   +   2-propanone   HO⊖ ⟶

   (c) Draw the most likely aldol product of the above reaction.

Model 3: Retrosynthetic Analysis of the Aldol Condensation

In retrosynthetic analysis (refer to CA13), the product is disconnected, or pulled apart, to determine the starting materials for the reaction.

Product   Synthons   Starting Materials

Questions:

6. Model 3 shows the partial retrosynthetic analysis of the aldol condensation.

   (a) Label the carbons of the product with Greek characters. At what position is the OH group? _____ Is this product an aldol condensation product? (*Circle one*) yes / no.

   (b) In an aldol condensation, what position is the new bond formed between (refer to #1f)? _____ On the product in Model 3, draw a line across the C-C bond that was formed to make the aldol product.

   (c) If the bond that was formed (marked in #6b) is disconnected or pulled apart, the synthons, or fragments, shown above are obtained. Complete these synthons by assigning a ⊕ charge to the fragment that acts as the E⊕ and a ⊖ charge to the fragment that acts as the Nu⊖.

   (d) On an aldehyde, which carbon is the Nu⊖ and which is the E⊕? Check to make sure this agrees with the charges placed on the synthons in #4c.

   (e) Finally, convert the synthons into actual reagents that could be used for the aldol condensation.

7. Determine the starting materials used to form each of the following aldol condensation products. (HINT: start by labeling carbons with Greek characters.)

   (a)

   (b)

Model 4: Dehydration of Aldol Products

Questions:

8. By applying heat, the aldol product undergoes dehydration as shown in Model 4.

   (a) Label the carbons of the aldol product and dehydration product with Greek characters.

   (b) Dehydration of the aldol product occurs with heat to give loss of _____.

   (c) The new double bond formed in the dehydration product occurs between which carbons? (Use Greek characters).

   (d) In the aldol product, circle the two groups that must be eliminated in order to form the dehydration product shown. (Draw in any hydrogens if necessary).

9. (a) In the following E2 elimination reaction, what two groups are eliminated in order to form the alkene? _____ Draw the mechanism to illustrate this mechanism.

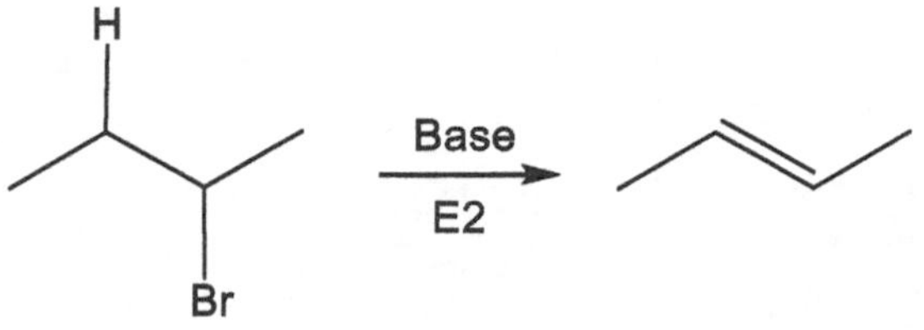

(b) Compare the elimination in #9a to the elimination to give the dehydration product shown below. Recall the two groups that are eliminated (#8d), then draw the mechanism to illustrate this E2 mechanism.

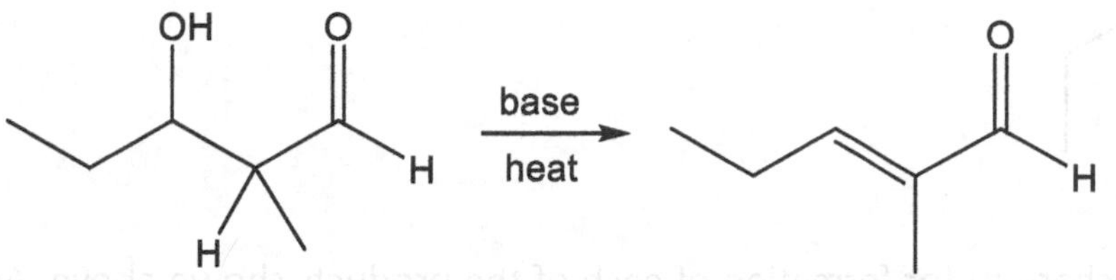

(c) Generally hydroxide groups (ⒽOH) are <u>not</u> good leaving groups. In your groups, discuss why this dehydration will occur even though the leaving group is poor.

Reflection: on a separate sheet of paper.
As a group, describe three concepts your group has learned from this activity and the one most important unanswered question about this activity that remains with your group. Turn this in before leaving class.

Additional Questions:
10. Draw the starting reagents used to prepare the following aldol products. Indicate whether the reaction is a self or a mixed condensation.
(a)

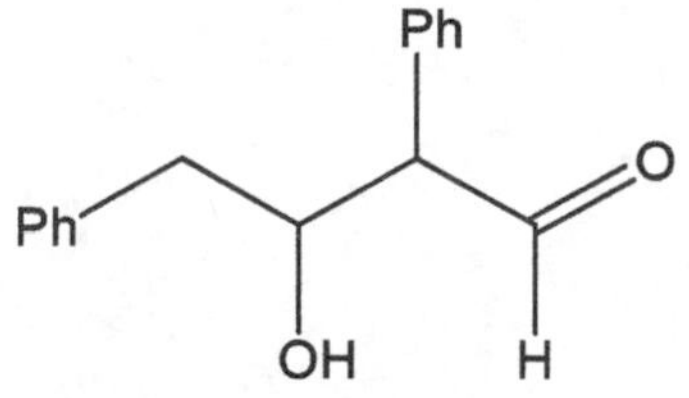

(b)

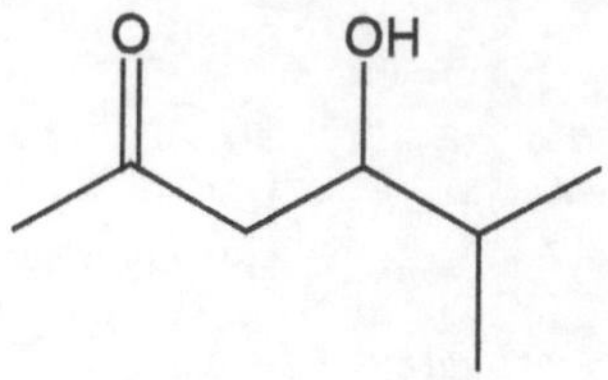

(c)

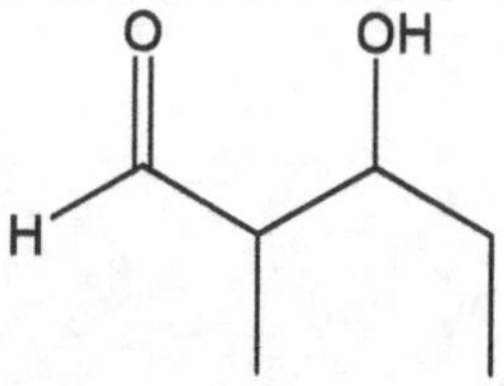

11. Draw the complete mechanism for formation of each of the products shown above. Assume the reagents are base catalyzed.

12. Aldol condensations occur in an intramolecular fashion to form five and six membered rings.
    (a)   Draw the product of the following intramolecular aldol condensation.

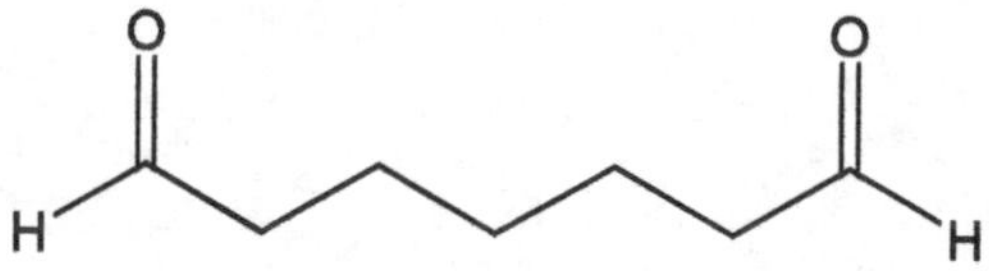

    (b)   Draw the starting material needed to form the following intramolecular aldol product.

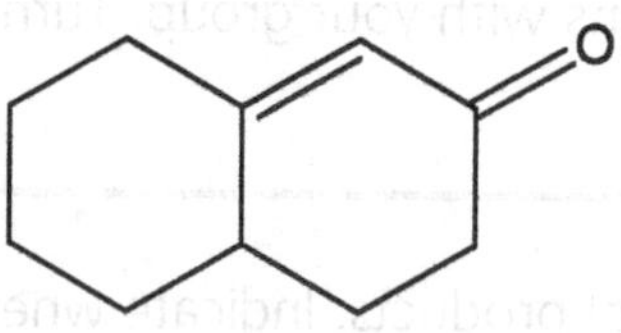

Class Activity 23C

Additional Reactions of Enolates

Prior Knowledge:
Before beginning this activity, students should be familiar with the following concepts:

- Acid, base, nucleophile and electrophile roles and properties.
- Formal charge, electronegativity, resonance, induction.
- Curved arrow notation.
- Enolate formation and aldol condensation.
- Decarboxylation.

Learning Objectives
Content Learning Objectives:
After completing this activity students should be able to:
- Predict the product and draw the mechanism for the Claisen condensation reaction (both self-condensation and mixed).
- Determine the most acidic hydrogen in a dicarbonyl compound and predict the reaction of the resulting enolate.
- Identify compounds that will act a Michael donors and Michael acceptors and predict the resulting products.

Process Objectives:
- Critical Thinking. Students evaluate using enolates of both esters and ketones in a variety of condensation type reactions.

Class Activity 23C

Additional Reactions of Enolates

Model 1: The Claisen Condensation

$$2 \quad \text{methyl acetate} \quad + \text{ NaOCH}_3 \longrightarrow \text{Claisen product}$$

Questions:

1. Consider the Clasien condensation shown in Model 1.
   (a) How many carbons are in the carbon framework of methyl acetate (not counting the OCH₃ group)? _____
   (b) How many carbons are in the Claisen product (not counting the OCH₃ group)? _____
   (c) How many molecules of methyl acetate combine to form the Claisen product? _____
   Circle each acetate portion in the Claisen product.
   (d) The Claisen product is also called a keto ester. At what position does the keto group reside with respect to the ester? (*Circle one*) α / β / γ / δ.
   (e) Using Greek characters, indicate the position where the new C-C bond occurs (two characters). _____
   (f) Compare the starting material to the product. What group has been eliminated?

2. (a) Recall the aldol condensation, shown below, from Class Activity 23B. Label the Nu⊖ and E⊕.

   How is the nucleophile generated?

   (b) Compare the aldol condensation to the Claisen condensation. How is the Claisen similar to the aldol condensation?

(c)   Draw the enolate ion that is formed from reacting methyl acetate with NaOCH₃. The enolate will react as the (circle one) Nu⊖ / E⊕.

3. Devise a mechanism for formation of the Claisen product, using curved arrows to show electron movement.

4. Like the aldol condensation, the Claisen can also afford mixed condensation products.
    (a)   Consider the reaction of an ester (pKa=24) with a ketone (pKa=20).

    Which compound is more acidic? (Circle one) ester / ketone.

    Which compound is most likely to act as the nucleophile in a mixed Claisen

    condensation? (Circle one) ester / ketone. Explain.

(b)   Draw the enolate formed from the ester and the ketone shown below. Explain why the ketone has a lower pKa than the ester.

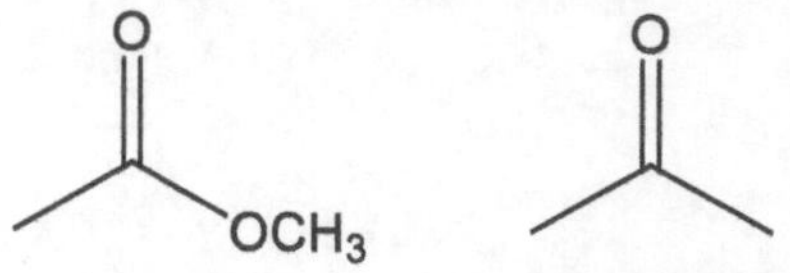

pKa= 24   pKa= 20

(c)   Draw the mixed Claisen condensation product that results from the base catalyzed reaction shown in #4b.

Model 2: Acetoacetic Ester Synthesis

acetoacetic ester

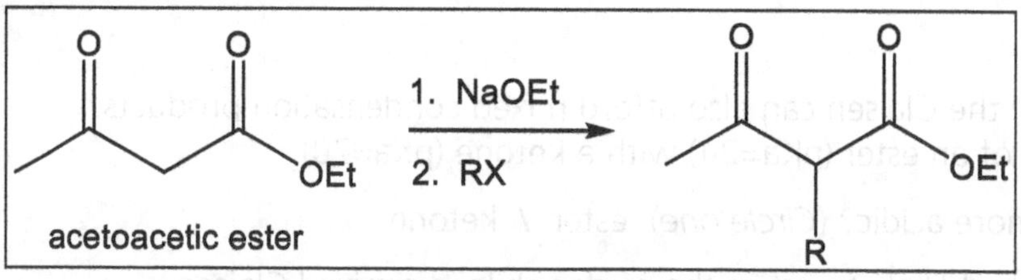

Questions:
5. Model 2 shows the acetoacetic ester synthesis.
    (a)   What are the two functional groups in acetoacetic ester?

          Label the carbons of acetoacetic ester with Greek characters, with respect to the ester group. At what position is the ketone group? _____ Could acetoacetic ester be prepared from a Claisen condensation (refer to Model 1)? (*Circle one*) yes / no.

(b)   How many acidic α–positions are present in acetoacetic ester? (*Circle one*) 1 / 2 / 3

(c)   Draw the enolate that would be formed from removal of the indicated hydrogen in the structures shown below. Draw all possible resonance forms for each enolate.

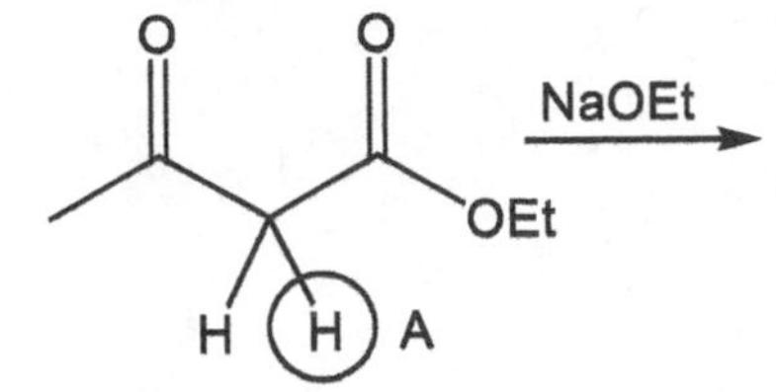

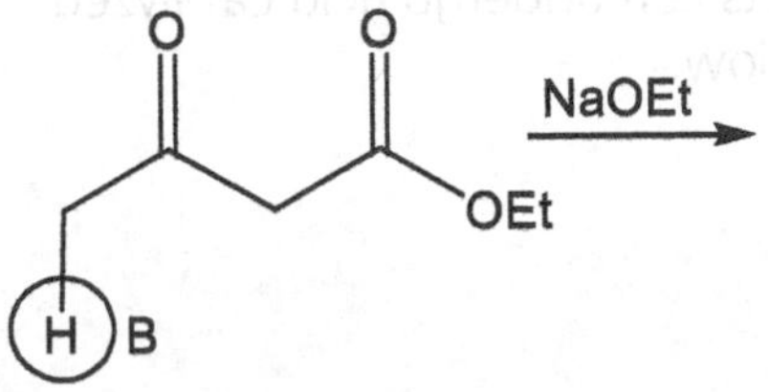

(d)   Which hydrogen, of the two choices shown in #5c, would be the most acidic? (*Circle one*) HA / HB. Explain.

(e)   In step 2, the enolate reacts with an alkyl halide, RX. The enolate will act as the (*circle one*) Nu⊖ / E⊕. Draw curved arrows to illustrate the mechanism for step 2.

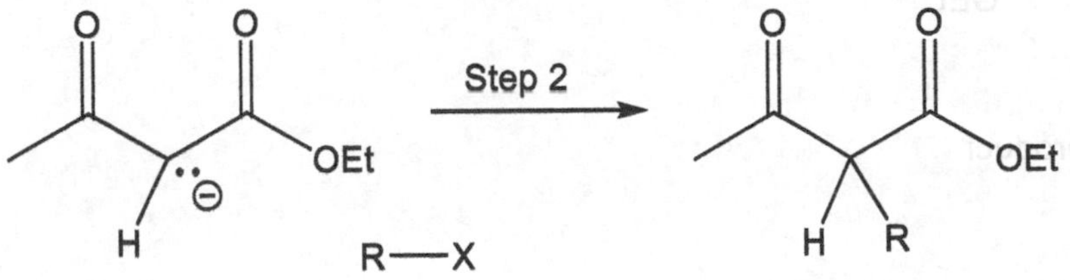

6. The malonic ester synthesis is shown below. Compare the malonic ester synthesis to the acetoacetic ester synthesis. Note any similarities and differences between these reactions.

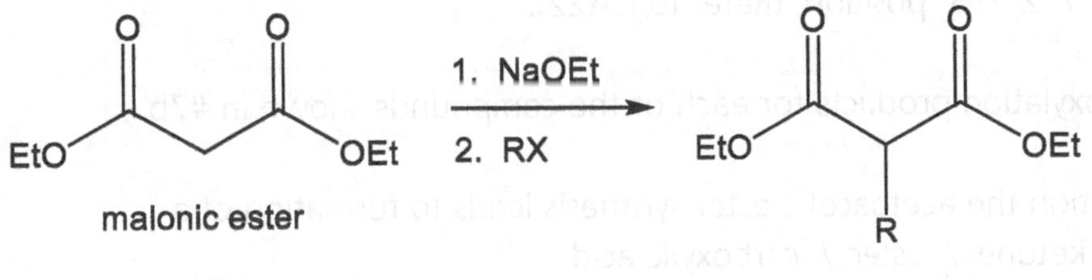

malonic ester

7. Esters can undergo acid catalyzed hydrolysis as shown below.

$$\text{CH}_3\text{C(=O)OCH}_3 + \text{H}_2\text{O} \underset{}{\overset{\text{H}^\oplus}{\rightleftharpoons}} \text{CH}_3\text{C(=O)OH} + \text{CH}_3\text{OH}$$

(a) What are the products of acid catalyzed hydrolysis?

(b) Acetoacetic ester and malonic ester synthesis products can undergo acid catalyzed hydrolysis. Draw the resulting hydrolysis products below.

acetoacetic ester product $\xrightarrow{\text{H}_3\text{O}^\oplus}$

malonic ester product $\xrightarrow{\text{H}_3\text{O}^\oplus}$

(c) Number each of the carboxylic acid products in #7b, where the carboxylic acid carbon is #1. The relationship between the carbonyl groups is (*circle one*) 1,2 / 1,3 / 1,4.

(d) Carboxylic acids can undergo decarboxylation when a second carbonyl group is located at the (*circle one*) 1 / 2 / 3 position. (Refer to CA22).

(e) Draw the decarboxylation products for each of the compounds shown in #7b.

After decarboxylation the acetoacetic ester synthesis leads to formation of a
     (*circle one*) ketone / ester / carboxylic acid.

After decarboxylation the malonic ester synthesis leads to formation of a
     (*circle one*) ketone / ester / carboxylic acid.

Model 3: Conjugate Addition (1, 4 - Addition) - the Michael Addition Reaction

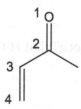

| α,β-unsaturated ketone | | 1,4-addition thermodynamic product | 1,2-addition kinetic product |

Questions:

8. Consider the conjugate addition (1, 4 - addition) reaction shown in Model 3.
   (a) What is an alternate name for the conjugate addition?

   (b) Explain why the starting material is called an α , β–unsaturated ketone.

   (c) The atoms involved in conjugate addition are numbered from 1-4 as shown in Model 3.
        What position number does the nucleophile add to form the 1,4-product? _____
        What position number does the nucleophile add to form the 1,2-product? _____

   (d) The thermodynamic product is the (circle one) 1,4-addition / 1,2-addition product.

        The thermodynamic product is the (circle one) more / less stable product.

9. (a) Draw two resonance forms for the α , β–unsaturated ketone shown below.

   (b) In the resonance structures drawn above, where do the positive charges occur (note the numbers)?

   (c) The α , β–unsaturated ketone acts as the (circle one) Nu⊖ / E⊕.

10. The choice of the nucleophile determines whether 1,4-addition or 1,2-addition occurs.
    (a)   Nucleophiles that are stabilized (less reactive) will afford 1,4-addition products. Circle the nucleophiles below that are considered stabilized.

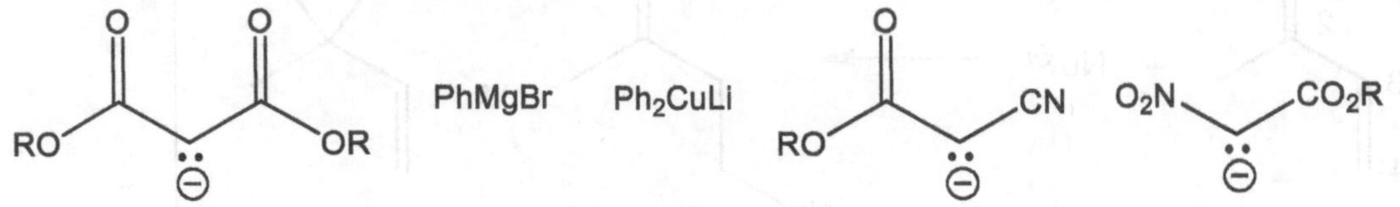

    (b)   Nucleophiles (circle one) donate / accept electrons. The nucleophile would act as the (circle one) Michael donor / Michael acceptor.

    (c)   Electrophiles (circle one) donate / accept electrons. The electrophile would act as the (circle one) Michael donor / Michael acceptor.

11. (a)   Label the starting materials in the reaction shown below as either the Michael donor or the Michael acceptor.

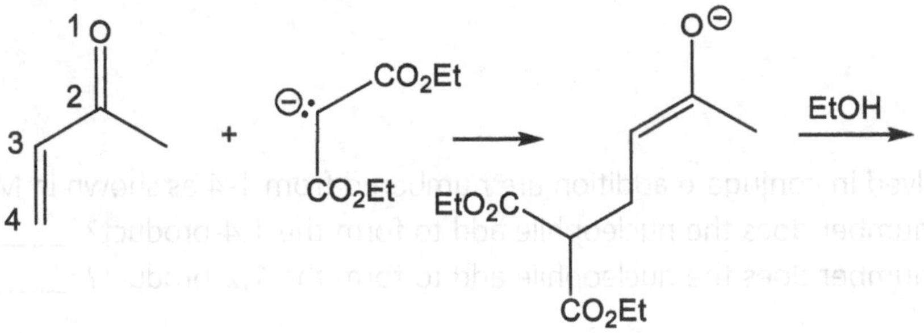

    (b)   How is malonic ester ($EtO_2CCH_2CO_2Et$) converted to the anion shown in #11a?

    (c)   Draw curved arrows to show electron movement for the Michael reaction above. Draw the final Michael addition product.

12. Draw the product for each of the following Michael addition reactions. (Recall that H_3O+ and heat result in hydrolysis and decarboxylation, see #7).
    (a)

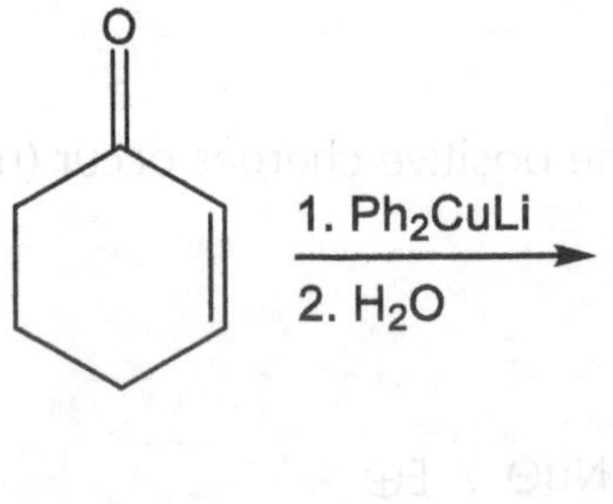

(b)

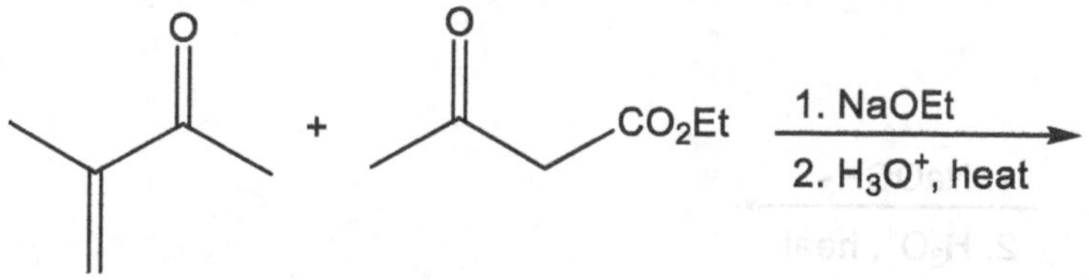

$$\text{methyl vinyl ketone} \; + \; \text{ethyl acetoacetate} \quad \xrightarrow[\text{2. } H_3O^+, \text{ heat}]{\text{1. NaOEt}}$$

Reflection: on a separate sheet of paper.

As a group, describe three concepts your group has learned from this activity and the one most important unanswered question about this activity that remains with your group. Turn this in before leaving class.

Additional Questions:

13. Draw the starting reagents used to prepare each of the following products.

(a)

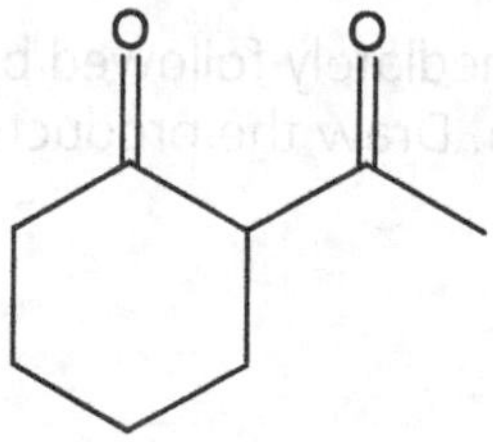

(b)

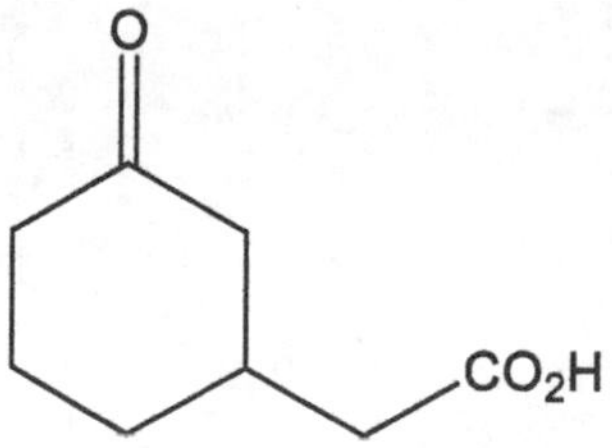

(c)   HINT: intramolecular reaction.

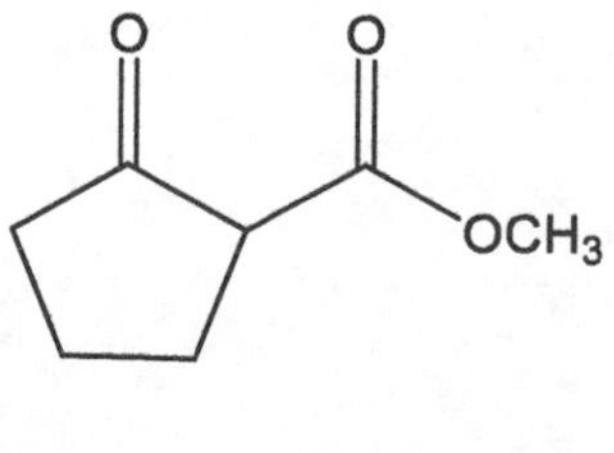

14. Draw the major organic product of the following sequence of reactions.

(a)

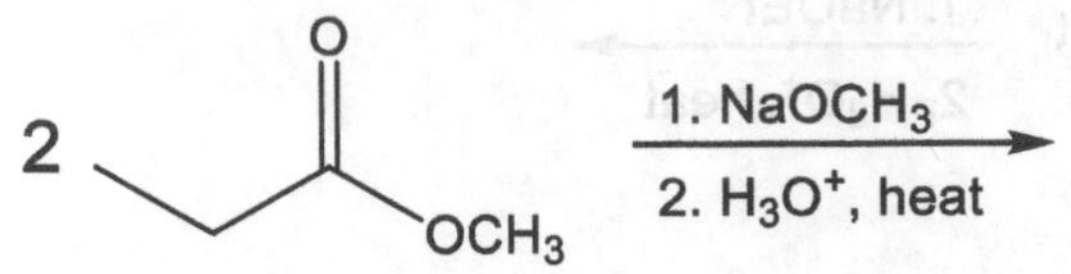

2
O
OCH3
1. NaOCH3
2. H3O+, heat

(b)

O
Ph
OEt
+
O
OEt
1. NaOEt (2 eq)
2.
O
3. H3O+, heat

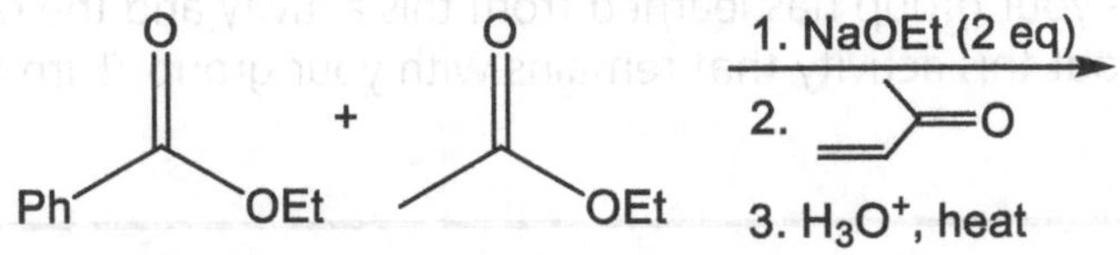

15. A ring forming reaction that occurs when a Michael addition is immediately followed by an intramolecular aldol condensation is called the Robinson annulation. Draw the products after each step of the following Robinson annulation reaction.

O
Ph
+
O
Ph
NaOH
heat

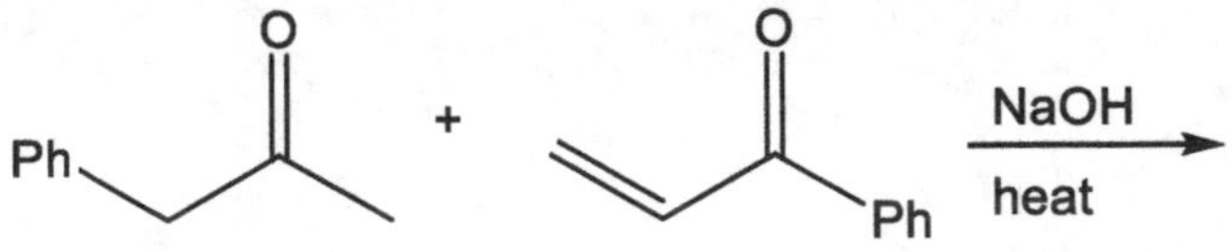

Class Activity 24

Reactions of Carboxylic Acid Derivatives

Prior Knowledge:

Before beginning this activity, students should be familiar with the following concepts:

- Acid, base, nucleophile and electrophile roles and properties.
- Resonance.
- Curved arrow notation.
- Substitution reactions and leaving group ability.

Learning Objectives

Content Learning Objectives:

After completing this activity students should be able to:
- Identify a carboxylic acid derivative and determine the relative reactivity as an electrophile.
- Relate reactivity of carboxylic acid derivatives to basicity of the leaving group.
- Predict the products and draw the mechanism of nucleophilic addition to carboxylic acid derivatives.

Process Objectives:

- Critical Thinking. Students analyze acid/base chemistry and leaving group ability in the context of carboxylic acid derivatives in order to predict reactivity and products for different reactions.

Class Activity 24

Reactions of Carboxylic Acid Derivatives

Model 1: Carboxylic Acid Derivative

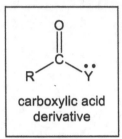

carboxylic acid
derivative

In the general structure of a carboxylic acid derivative, "Y" is any atom that contains at least one lone pair of electrons.

Questions:
1. Consider the carboxylic acid derivative shown in Model 1.
   (a) Draw two additional resonance forms for the carboxylic acid derivative.

   (b) Draw the resonance forms for a ketone. Compare the resonance forms of the carboxylic acid derivative to the resonance forms of a ketone. How are they different?

   (c) Which atom in the carboxylic acid derivative would act as the electrophile? Explain.

   (d) Consider the "Y" group on the carboxylic acid derivative. A "Y" group that (*circle one*) donates / accepts electrons would make the acid derivative a <u>more</u> reactive electrophile. Explain.

(e) A "Y" group that (*circle one*) donates / accepts electrons would make the acid derivative a **less** reactive electrophile. Explain.

2. Identify which of the following are carboxylic acid derivatives by circling the "Y" group in each compound. Put an X through any compounds that are not carboxylic acid derivatives.

Model 2: Nucleophilic Substitution of Carboxylic Acid Derivatives

Questions:

3. Consider the reaction shown in Model 2.

(a) What is the role of the carboxylic acid derivative?

(*Circle one*) nucleophile / electrophile

(b) A nucleophile reacts with a carboxylic acid derivative. What is the leaving group?

(c) If Y⊖ is a strong base, the reaction would proceed in the (*circle one*) forward / reverse direction. Explain.

(d) If Y⊖ is a weak base, the reaction would proceed in the (*circle one*) forward / reverse direction. Explain.

(e) In your groups discuss how the basicity of Y⊖ affects the reactivity of the acid derivative.

4. In the following table:

Acid Derivative	Leaving Group YΘ	pKa	Reactivity
RCOCl		-7	
RCO₂COR		3-5	
RCO₂R'		16-18	
RCONR'₂		38-40	

(a) Draw the appropriate leaving group, YΘ, for each acid derivative in the table.

(b) Based on the pKa values given, which Y group is most basic? Which Y group is least basic?

(c) Rank the acid derivatives in the table above, from 1 to 4, where 1 is the most reactive and 4 is the least reactive.

(d) Once your group has reached agreement, discuss whether your conclusions (for the table) agree with the answer to #3e above.

Model 3: Nucleophilic Acyl Substitution

In the reaction of carboxylic acid derivatives:
- The nucleophile can be negative or neutral.
- Substitution occurs because Y is a leaving group.
- The mechanism always occurs via a tetrahedral intermediate

tetrahedral intermediate

Questions:
5. For the reaction shown in Model 3:
   (a) Draw curved arrows to show electron movement for both steps of the mechanism.

(b)   Why is the intermediate called the tetrahedral intermediate?

What other similar reaction also forms a tetrahedral intermediate?

(c)   How does leaving group ability affect the reactivity of the carboxylic acid derivative?

(d)   From the table shown in problem #4 above, rank the leaving groups from best leaving group (1) to poorest leaving group (4).

(e)   How do pKa values relate to leaving group ability?

(f)   Does your ranking of leaving groups agree with your ranking of reactivity of acid derivatives in the table from question #4? Explain why or why not.

6. Complete the following scheme by providing the nucleophile needed to accomplish each reaction. (Draw the nucleophile above the arrow).

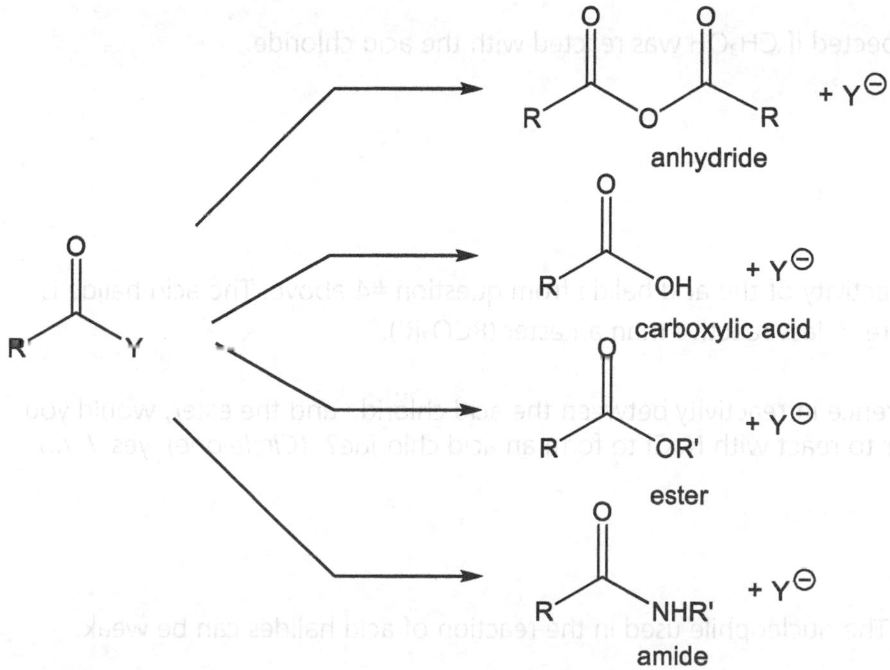

Model 4: Reaction of Acid Halides

Acid halides react with a variety of nucleophiles, forming HCl as a byproduct. A weak base such as pyridine is added in order to neutralize the HCl formed.

Questions:

7. For the reaction shown in Model 4:

   (a) Identify the nucleophile and the electrophile for the first step.

   (b) The nucleophile is (*circle one*) strong / weak.

   (c) Draw the structure of the tetrahedral intermediate that is formed after the first step.

   (d) Draw curved arrows to show electron movement from the first step through the tetrahedral intermediate, then to give the product. (Refer to Model 3).

   (e) HCl is formed in this reaction. Discuss the purpose of adding a weak base like pyridine to the reaction mixture.

8. Draw the product expected if CH₃OH was reacted with the acid chloride.

9. (a) Consider the reactivity of the acid halide from question #4 above. The acid halide is (*circle one*) more / less reactive than an ester (RCO₂R′).

   (b) Given the difference in reactivity between the acid chloride and the ester, would you expect the ester to react with NaCl to form an acid chloride? (*Circle one*) yes / no. Explain.

   (c) **True or False.** The nucleophile used in the reaction of acid halides can be weak.

(d)   Given the reactivity of the ester, compared to the acid halide, discuss whether the ester would be reactive with a weak nucleophile such as water.

(e)   If the ester will not react with water, what reagent could be added to enhance this reaction? Explain the purpose of this reagent.

Reflection: on a separate sheet of paper.
As a group, describe three concepts your group has learned from this activity and the one most important unanswered question about this activity that remains with your group. Turn this in before leaving class.

Additional Questions:
10. (a)   Draw the products that would result on reaction of each of the acid derivatives with the reagents shown in the table below.

Acid derivative → + Reagent ↓	$Ph-C(=O)-OCH_3$	$CH_3CH_2-C(=O)-Cl$	$cyclopropyl-C(=O)-NH_2$	$CH_3-C(=O)-O-C(=O)-CH_3$
H_2O				
CH_3CH_2OH				
CH_3NH_2				

(b)   For each of the above reactions, determine if an acid catalyst would be necessary.

11. A mixed anhydride, shown below, is reacted with excess methanol, under acidic conditions. Draw the two organic products that are formed.

12. Fatty acids can flow through the cell membrane to enter fat cells in the body. Once in the fat cells, the fatty acids can react with glycerol to form triglycerides, which are stored in the fat cells. Draw the structure of the triglyceride formed in this reaction, assuming excess fatty acid is present.

fatty acid
R=>12 carbons

+

glycerol

enzyme →

Class Activity 25

Amines

Prior Knowledge:
Before beginning this activity, students should be familiar with the following concepts:

- Acid, base, nucleophile and electrophile roles and properties.
- Stereochemistry
- Curved arrow notation.
- Newman projections.
- Elimination (E1 and E2 mechanisms)

Learning Objectives
Content Learning Objectives:
After completing this activity students should be able to:
- Predict the relative basicity of different substituted amines.
- Predict the product of an acid base reaction with an amine.
- Draw the products expected from alkylation and Hoffman elimination of amine.

Process Objectives:
- Critical Thinking. Students analyze and evaluate base strength of amines and the reactions of amines as both bases and nucleophiles in a variety of different reactions.

Class Activity 25

Amines

Model 1: Basicity of Amines

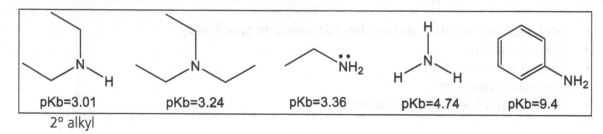

| pKb=3.01 | pKb=3.24 | pKb=3.36 | pKb=4.74 | pKb=9.4 |
| 2° alkyl | | | | |

Questions:

1. Consider the amines shown in Model 1. The first amine on the left is labeled as 2° alkyl.
   (a) For this 2° alkyl amine, how many R groups are attached to the nitrogen? _____ These R groups are (*circle one*) alkyl groups / aromatic groups.
   (b) Identify the remaining amines as alkyl or aryl.
   (c) Identify the remaining amines as 1°, 2° or 3°.

2. A low pKa value indicates the compound is a (*circle one*) strong / weak acid.
   A low pKb value indicates the compound is a (*circle one*) strong / weak base.

3. From the information in Model 1, identify which compound is the strongest base and which compound is the weakest base.

4. Using the general classification of amines (ammonia, 1°, 2° or 3°), rank from strongest to weakest using the data in Model 1.

Model 2: Amines as Bases

$$CH_3NH_2 \;+\; H\text{-}O\text{-}H \longrightarrow CH_3\overset{\oplus}{NH_3} \;+\; HO^{\ominus}$$

$$(CH_3)_2NH \;+\; H\text{-}O\text{-}H \longrightarrow$$

Questions:

5. In the reactions in Model 2, the amine acts as the base.
   (a) A base (*circle one*) donates / accepts electrons.
       A base (*circle one*) donates / accepts a proton.

(b)   In the reaction of methylamine (top reaction in Model 2), identify the acid, base, and conjugate acid/base. Draw curved arrows to indicate electron flow.

(c)   Draw the products that result when <u>dimethylamine</u> (bottom reaction in Model 2) reacts with water.

(d)   In the above reactions, the amine base reacts with water to form the conjugate acid. The conjugate acid has a (*circle one*) positive / negative charge. In general what types of effects will help stabilize the conjugate acid?

(e)   Compare the conjugate acid formed from methyl amine (1°) and the conjugate acid formed from dimethylamine (2°). Recall that alkyl groups donate electrons by induction. Which conjugate acid is more stabilized by induction? Explain. Draw polarity arrows on the C-N bond to show this inductive effect.

(f)   Draw the conjugate acid formed when <u>trimethylamine</u> reacts with water. Is this product more or less stabilized by induction compared to the 1° and 2° amines shown above?

(g)   If the conjugate acid is more stable, that would indicate that the base is more basic. Look at the pKb values given in Model 1. Is the 3° amine the most basic?

6. A second effect that helps stabilize the conjugate acid is called solvation, which occurs via hydrogen bonding between the conjugate acid and the lone pairs on the oxygen of water.

(a)   Show the hydrogen bonding that occurs with the conjugate acid from a 3° amine (with H_2O) and the conjugate acid of a 2° amine (with H_2O).

(b)   How does the H-bonding change the polarity of the N-H bond? Draw polarity arrows on
       the NH bond to show this effect.

(c)   Once your group agrees on the above, discuss whether the solvation effect stabilizes or
       destabilizes the conjugate acid.

7. (a)   Draw the conjugate acid for each of the following bases.

(b)   Rationalize why an aromatic amine is a weaker base than an alkyl amine.

(c)   Rationalize why the amide is an even weaker base than an aryl amine.

Model 3: Alkylation of Amines

• Exhaustive Alkylation

NH_3 + Br—CH_3 $\xrightarrow{\text{Rxn 1}}$ CH_3NH_2 $\xrightarrow{\text{Rxn 2}}$ $(CH_3)_2NH$ $\xrightarrow{\text{Rxn 3}}$ $(CH_3)_3N$ $\xrightarrow{\text{Rxn 4}}$ $(CH_3)_4N^{\oplus}$
(excess) $Br^{\ominus}$

• Selective Alkylation (Gabrielle Synthesis)

[structures] $\xrightarrow[\text{2. }CH_3Br]{\text{1. base}}$ [structure] $\xrightarrow[\text{heat}]{H_2NNH_2}$ [structure] + $CH_3\overset{\oplus}{N}H_3$ → CH_3NH_2

only monoalkylated
1° amine formed

Questions:

8. Consider the exhaustive alkylation shown in Model 3.
   (a) For Rxn 1, label the starting reagents as $Nu^{\ominus}/E^{\oplus}$ and draw curved arrows to show electron movement.
   (b) Compare the product formed in Rxn 1 to the starting material. Which compound would be the more reactive nucleophile? (Circle one) NH_3 / CH_3NH_2 Explain.

   (c) Would you expect the reaction of ammonia to stop after formation of the primary amine? (Circle one) yes / no.

   (d) Explain why the rate of the four reactions gets faster as the reaction progresses from Rxn 1 to Rxn 4.

   (e) As a group discuss why when alkylating 1° amines, only a little 2° amine product is formed.

9. Consider the selective alkylation shown in Model 3.
   (a) What is the final product that results from this selective alkylation?

   (b) Can this final product be readily formed via the alkylation method shown in exhaustive alkylation procedure?

10. Alkyl amines are generally very weak acids. It takes a large amount of energy to pull off a H from an amine to form a nitrogen anion. Explain why only a weak base is necessary to form the N anion shown in the first step of the selective alkylation.

Model 4: Hoffman Elimination

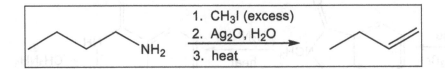

Questions:

11. (a)   Step 1 of the Hoffman Elimination is called exhaustive methylation, which occurs in excess methyl iodide. Draw the product of this over-alkylation (see Model 3).

    (b)   Step 2 involves exchanging the anion from I⊖ to HO⊖. Draw the product.

    (c)   Step 3 is β- elimination which occurs on heating. Elimination is driven by the basic counterion (HO⊖). Label the α and β–positions.

    (d)   Draw the product of step 2 in a Newman projection along the α–β bond. For β–elimination, what relationship is needed between the leaving group and the β–H? Rotate the Newman projection, if necessary, so that the elimination can occur.

12. (a)   The following compound is formed after the second step of the Hoffman Elimination. Identify the two types of β–hydrogens, then draw the two possible elimination products.

(b)   Recall that the most substituted alkene is the Zaitsev product while the less substituted alkene is the Hoffman product. Label each product above as Zaitsev or Hoffman.

(c)   Under Hoffman conditions, the Hoffman product is the major elimination product. Draw a Newman projection along C2-C3. Rotate so the NMe₃ and the β–H are anti. Would you consider this to be a stable conformation? Explain.

(d)   Draw a Newman projection along the C1-C2 bond. Rotate so the NMe₃ and the β–H are anti. Would you consider this to be a stable conformation? Explain.

(e)   As a group discuss the reason why the Hoffman product is favored over the Zaitsev product, even though the Zaitsev is the more stable product.

Reflection: on a separate sheet of paper.

As a group, describe three concepts your group has learned from this activity and the one most important unanswered question about this activity that remains with your group. Turn this in before leaving class.

Additional Questions:

13. Draw the conjugate acid formed from each amine base. For each pair of amines, determine which amine is the most basic. Explain.

(a)

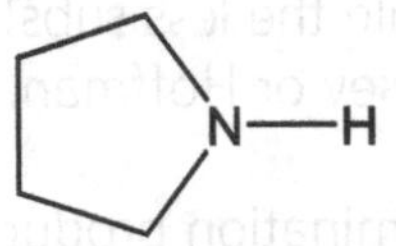

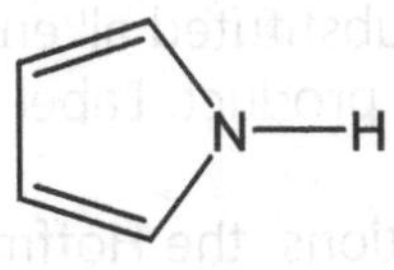

(b)

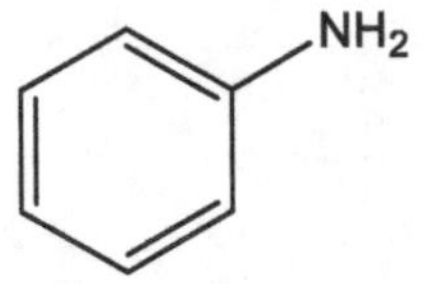

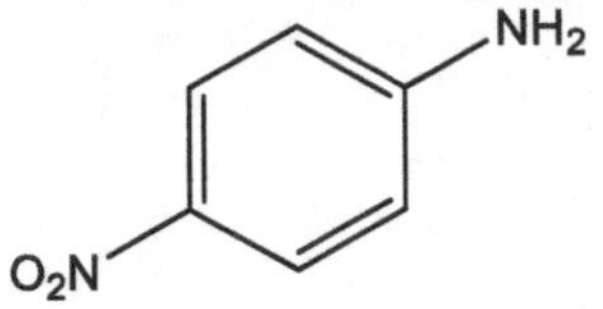

(c)

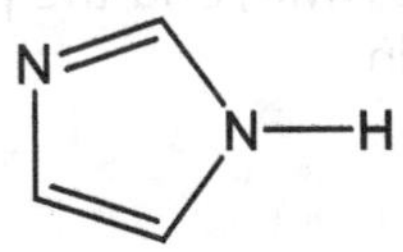

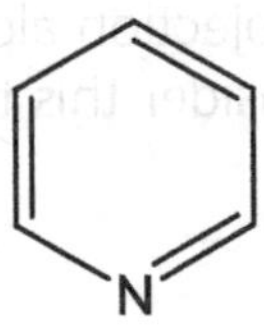

14. Rank the following compounds from most basic to least basic. Explain.

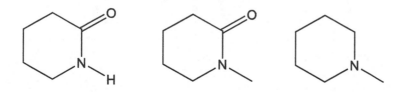

15. Draw the product for each acid-base reaction below. Determine if the reaction will proceed in the forward or the reverse direction.

(a)

N—H + H_2O ⟶

(b)

N—H + CH_3CO_2H ⟶

16. Determine the products of the following sequence of reactions.

NH	1. NaOtBu		1. CH_3I (xs)	
	2. $CH_3CH_2CH_2Br$		2. Ag_2O, H_2O, heat	
	3. H_2NNH_2, heat		3. BH_3, H_2O_2; then NaOH	
			4. PCC	

14. Rank the following compounds from most basic to least basic. Explain.

15. Draw the product for each acid-base reaction below. Determine if the reaction will proceed in the forward or reverse direction.

(a)

$$\text{N–H} + H_2O \longrightarrow$$

(b)

$$\text{N–H} + CH_3CO_2H \longrightarrow$$

16. Determine the products of the following sequence of reactions.

1. NaOBu
2. $CH_3CH_2CH_2Br$
3. H_2NNH_2, heat

1. CH_3I (xs)
2. Ag_2O, H_2O, heat
3. BH_3; H_2O_2, then NaOH
4. PCC